# TOWN GEOLOGY

## CHARLES KINGSLEY

**Serenity**
Publishers, LLC
ROCKVILLE, MARYLAND

2009

ISBN:   978-1-60450-610-5

Published by Serenity Publishers
An Arc Manor Company
P. O. Box 10339
Rockville, MD 20849-0339
www.SerenityPublishers.com

Printed in the United States of America/United Kingdom

# CONTENTS

PREFACE                                5

I.   THE SOIL OF THE FIELD             25

II.  THE PEBBLES IN THE STREET         39

III. THE STONES IN THE WALL            57

IV.  THE COAL IN THE FIRE              72

V.   THE LIME IN THE MORTAR            89

VI.  THE SLATES ON THE ROOF            100

# PREFACE

This little book, including the greater part of this Preface, has shaped itself out of lectures given to the young men of the city of Chester. But it does not deal, in its present form, with the geology of the neighbourhood of Chester only. I have tried so to recast it, that any townsman, at least in the manufacturing districts of England and Scotland, may learn from it to judge, roughly perhaps, but on the whole accurately, of the rocks and soils of his own neighbourhood. He will find, it is true, in these pages, little or nothing about those "Old Red Sandstones," so interesting to a Scotchman; and he will have to bear in mind, if he belong to the coal districts of Scotland, that the "stones in the wall" there belong to much older rocks than those "New Red Sandstones" of which this book treats; and that the coal measures of Scotland, with the volcanic rocks which have disturbed them, are often very different in appearance to the English coal measures. But he will soon learn to distinguish the relative age of rocks by the fossils found in them, which he can now, happily, study in many local museums; and he may be certain, for the rest, that all rocks and soils whatsoever which he may meet have been laid down by the agents, and according to the laws, which I have tried to set forth in this book; and these only require, for the learning of them, the exercise of his own observation and common sense. I have not tried to make this a handbook of geological facts. Such a guide (and none better) the young man will find in Sir Charles Lyell's "Student's Elements of Geology." I have tried rather to teach the method of geology, than its facts; to furnish the student with a key to all geology, rough indeed and rudimentary, but sure and sound enough, I trust, to help him to unlock most geological problems which he may meet, in any quarter of the globe. But young men must remember always, that neither this book, nor all the books in the world, will make them geologists. No amount of book learning will make a man a scientific man; nothing but patient observation, and quiet and fair thought over what he has observed. He must go out for himself, see for himself, compare and judge for himself, in the field, the quarry, the cutting. He must study rocks, ores, fossils, in the nearest museum; and thus store his head, not with words, but with facts. He must verify—as far as he can—what he reads in books, by his own observation; and be slow to believe anything, even on the highest scientific authority, till he has either seen it, or something like enough to it to make it seem to him probable, or at least possible. So, and so only, will he become a scientific man, and a good geologist; and acquire that habit of mind by which alone he can judge fairly and wisely of facts of any kind whatsoever.

I say—facts of any kind whatsoever. If any of my readers should be inclined to say to themselves: Geology may be a very pleasant study, but I have no special fancy for it. I had rather learn something of botany, astronomy, chemistry, or what not—I shall answer: By all means. Learn any branch of Natural Science you will. It matters little to me which you learn, provided you learn one at least. But bear in mind, and settle it in your hearts, that you will learn no branch of science soundly, so as to master it, and be able to make use of it, unless you acquire that habit and method of mind which I am trying to teach you in this book. I have tried to teach it you by geology, because geology is, perhaps, the simplest and the easiest of all physical sciences. It appeals more than any to mere common sense. It requires fewer difficult experiments, and expensive apparatus. It requires less previous knowledge of other sciences, whether pure or mixed; at least in its rudimentary stages. It is more free from long and puzzling Greek and Latin words. It is specially, the poor man's science. But if you do not like it, study something else. Only study that as you must study geology; proceeding from the known to the unknown by observation and experiment.

But here some of my readers may ask, as they have a perfect right to ask, why I wish young men to learn Natural Science at all? What good will the right understanding of geology, or of astronomy, or of chemistry, or of the plants or animals which they meet—what good, I say, will that do them?

In the first place, they need, I presume, occupation after their hours of work. If any of them answer: "We do not want occupation, we want amusement. Work is very dull, and we want something which will excite our fancy, imagination, sense of humour. We want poetry, fiction, even a good laugh or a game of play"—I shall most fully agree with them. There is often no better medicine for a hard-worked body and mind than a good laugh; and the man who can play most heartily when he has a chance of playing is generally the man who can work most heartily when he must work. But there is certainly nothing in the study of physical science to interfere with genial hilarity; though, indeed, some solemn persons have been wont to reprove the members of the British Association, and specially that Red Lion Club, where all the philosophers are expected to lash their tails and roar, of being somewhat too fond of mere and sheer fun, after the abstruse papers of the day are read and discussed. And as for harmless amusement, and still more for the free exercise of the fancy and the imagination, I know few studies to compare with Natural History; with the search for the most beautiful and curious productions of Nature amid her loveliest scenery, and in her freshest atmosphere. I have known again and again working men who in the midst of smoky cities have kept their bodies, their minds, and their hearts healthy and pure by going out into the country at odd hours, and making collections of fossils, plants, insects, birds, or some other objects of natural history; and I doubt not that such will be the case with some of my readers.

Another argument, and a very strong one, in favour of studying some branch of Natural Science just now is this—that without it you can hardly keep pace with the thought of the world around you.

Over and above the solid gain of a scientific habit of mind, of which I shall speak presently, the gain of mere facts, the increased knowledge of this planet on which we live, is very valuable just now; valuable certainly to all who do not wish their children and their younger brothers to know more about the universe than they do.

Natural Science is now occupying a more and more important place in education. Oxford, Cambridge, the London University, the public schools, one after another, are taking up the subject in earnest; so are the middle-class schools; so I trust will all primary schools throughout the country; and I hope that my children, at least, if not I myself, will see the day, when ignorance of the primary laws and facts of science will be looked on as a defect, only second to ignorance of the primary laws of religion and morality.

I speak strongly, but deliberately. It does seem to me strange, to use the mildest word, that people whose destiny it is to live, even for a few short years, on this planet which we call the earth, and who do not at all intend to live on it as hermits, shutting themselves up in cells, and looking on death as an escape and a deliverance, but intend to live as comfortably and wholesomely as they can, they and their children after them— it seems strange, I say, that such people should in general be so careless about the constitution of this same planet, and of the laws and facts on which depend, not merely their comfort and their wealth, but their health and their very lives, and the health and the lives of their children and descendants.

I know some will say, at least to themselves: "What need for us to study science? There are plenty to do that already; and we shall be sure sooner or later to profit by their discoveries; and meanwhile it is not science which is needed to make mankind thrive, but simple common sense."

I should reply, that to expect to profit by other men's discoveries when you do not pay for them—to let others labour in the hope of entering into their labours, is not a very noble or generous state of mind—comparable somewhat, I should say, to that of the fatting ox, who willingly allows the farmer to house him, till for him, feed him, provided only he himself may lounge in his stall, and eat, and *not* be thankful. There is one difference in the two cases, but only one—that while the farmer can repay himself by eating the ox, the scientific man cannot repay himself by eating you; and so never gets paid, in most cases, at all.

But as for mankind thriving by common sense: they have not thriven by common sense, because they have not used their common sense according to that regulated method which is called science. In no age, in no country, as yet, have the majority of mankind been guided, I will not say by the love of God, and by the fear of God, but even by sense and reason. Not sense and reason, but nonsense and unreason, prejudice and fancy, greed and haste, have led them to such results as were to be expected—to superstitions, persecutions, wars, famines, pestilence, hereditary diseases, poverty, waste—waste incalculable, and now too often irremediable—waste of life, of labour, of capital, of raw material, of soil, of manure, of every bounty which God has bestowed on man, till, as in the eastern Mediterranean, whole countries, some of the finest in the world, seem ruined for ever: and all because men will not learn nor obey those physical laws of the universe, which (whether we be conscious of them or not) are all around us, like walls of iron and of adamant—say rather, like some vast machine, ruthless though beneficent, among the wheels of which if we entangle ourselves in our rash ignorance, they will not stop to set us free, but crush us, as they have crushed whole nations and whole races ere now, to powder. Very terrible, though very calm, is outraged Nature.

> *Though the mills of God grind slowly,*
> *Yet they grind exceeding small;*
> *Though He sit, and wait with patience,*
> *With exactness grinds He all.*

It is, I believe, one of the most hopeful among the many hopeful signs of the times, that the civilised nations of Europe and

America are awakening slowly but surely to this truth. The civilised world is learning, thank God, more and more of the importance of physical science; year by year, thank God, it is learning to live more and more according to those laws of physical science, which are, as the great Lord Bacon said of old, none other than "Vox Dei in rebus revelata"—the Word of God revealed in facts; and it is gaining by so doing, year by year, more and more of health and wealth; of peaceful and comfortable, even of graceful and elevating, means of life for fresh millions.

If you want to know what the study of physical science has done for man, look, as a single instance, at the science of Sanatory Reform; the science which does not merely try to cure disease, and shut the stable-door after the horse is stolen, but tries to prevent disease; and, thank God! is succeeding beyond our highest expectations. Or look at the actual fresh amount of employment, of subsistence, which science has, during the last century, given to men; and judge for yourselves whether the study of it be not one worthy of those who wish to help themselves, and, in so doing, to help their fellow-men. Let me quote to you a passage from an essay urging the institution of schools of physical science for artisans, which says all I wish to say and more:

"The discoveries of Voltaic electricity, electromagnetism, and magnetic electricity, by Volta, Œrsted, and Faraday, led to the invention of electric telegraphy by Wheatstone and others, and to the great manufactures of telegraph cables and telegraph wire, and of the materials required for them. The value of the cargo of the Great Eastern alone in the recent Bombay telegraph expedition was calculated at three millions of pounds sterling. It also led to the employment of thousands of operators to transmit the telegraphic messages, and to a great increase of our commerce in nearly all its branches by the more rapid means of communication. The discovery of Voltaic electricity further led to the invention of electro-plating, and to the employment of a large number of persons in that business. The numerous experimental researches on specific heat, latent heat, the tension of vapours, the properties of water, the mechanical effect of heat, etc., resulted in the development of steam-engines, and railways, and the almost endless employments depending upon their construction and use. About a quarter of a million of persons are employed on railways alone in Great Britain. The various original investigations on the chemical effects of light led to the invention of photography, and have given employment to thousands of persons who practise that process, or manufacture and prepare the various material and articles required in it. The discovery of chlorine by Scheele led to the invention of the modern processes of bleaching, and to various improvements in the dyeing of the textile fabrics, and has given employment to a very large number of our Lancashire operatives. The discovery of chlorine has also contributed to the employment of thousands of printers, by enabling Esparto grass to be bleached and formed into paper for the use of our daily press. The numerous experimental investigations in relation to coal-gas have been the means of extending the use of that substance, and of increasing the employment of workmen and others connected with its manufacture. The discovery of the alkaline metals by Davy, of cyanide of potassium, of nickel, phosphorus, the common acids, and a multitude of other substances, has led to the employment of a whole army of workmen in the conversion of those substances into articles of utility. The foregoing examples might be greatly enlarged upon, and a great many others might be selected from the sciences of physics and chemistry: but those mentioned will suffice. There is not a force of Nature, nor scarcely a material substance that we employ, which has not been the subject of several, and in some cases of numerous, original experimental researches, many of which have resulted, in a greater or less degree, in increasing the employment for workmen and others." [1]

"All this may be very true. But of what practical use will physical science be to me?"

Let me ask in return: Are none of you going to emigrate? If you have courage and wisdom, emigrate you will, some of you, instead of stopping here to scramble over each other's backs for the scraps, like black-beetles in a kitchen. And if you emigrate,

---

1     See "Nature," No. XXV. (Macmillan & Co.)

you will soon find out, if you have eyes and common sense, that the vegetable wealth of the world is no more exhausted than its mineral wealth. Exhausted? Not half of it—I believe not a tenth of it—is yet known. Could I show you the wealth which I have seen in a single Tropic island, not sixty miles square—precious timbers, gums, fruits, what not, enough to give employment and wealth to thousands and tens of thousands, wasting for want of being known and worked—then you would see what a man who emigrates may do, by a little sound knowledge of botany alone.

And if not. Suppose that any one of you, learning a little sound Natural History, should abide here in Britain to your life's end, and observe nothing but the hedgerow plants, he would find that there is much more to be seen in those mere hedgerow plants than he fancies now. The microscope will reveal to him in the tissues of any wood, of any seed, wonders which will first amuse him, then puzzle him, and at last (I hope) awe him, as he perceives that smallness of size interferes in no way with perfection of development, and that "Nature," as has been well said, "is greatest in that which is least." And more. Suppose that he went further still. Suppose that he extended his researches somewhat to those minuter vegetable forms, the mosses, fungi, lichens; suppose that he went a little further still, and tried what the microscope would show him in any stagnant pool, whether fresh water or salt, of Desmidiæ, Diatoms, and all those wondrous atomies which seem as yet to defy our classification into plants or animals. Suppose he learnt something of this, but nothing of aught else. Would he have gained no solid wisdom? He would be a stupider man than I have a right to believe any of my readers to be, if he had not gained thereby somewhat of the most valuable of treasures—namely, that inductive habit of mind, that power of judging fairly of facts, without which no good or lasting work will be done, whether in physical science, in social science, in politics, in philosophy, in philology, or in history.

But more: let me urge you to study Natural Science, on grounds which may be to you new and unexpected—on social, I had almost said on political, grounds.

We all know, and I trust we all love, the names of Liberty, Equality, and Brotherhood. We feel, I trust, that these words are too beautiful not to represent true and just ideas; and that therefore they will come true, and be fulfilled, somewhen, somewhere, somehow. It may be in a shape very different from that which you, or I, or any man expects; but still they will be fulfilled.

But if they are to come true, it is we, the individual men, who must help them to come true for the whole world, by practising them ourselves, when and where we can. And I tell you—that in becoming scientific men, in studying science and acquiring the scientific habit of mind, you will find yourselves enjoying a freedom, an equality, a brotherhood, such as you will not find elsewhere just now.

Freedom: what do we want freedom for? For this, at least; that we may be each and all able to think what we choose; and to say what we choose also, provided we do not say it rudely or violently, so as to provoke a breach of the peace. That last was Mr. Buckle's definition of freedom of speech. That was the only limit to it which he would allow; and I think that that is Mr. John Stuart Mill's limit also. It is mine. And I think we have that kind of freedom in these islands as perfectly as any men are likely to have it on this earth.

But what I complain of is, that when men have got the freedom, three out of four of them will not use it. What?—someone will answer—Do you suppose that I will not say what I choose, and that I dare not speak my own mind to any man? Doubtless. But are you sure first, that you think what you choose, or only what someone else chooses for you? Are you sure that you make up your own mind before you speak, or let someone else make it up for you? Your speech may be free enough, my good friend; and Heaven forbid that it should be anything else: but are your thoughts free likewise? Are you sure that, though you may hate bigotry in others, you are not somewhat of a bigot yourself? That you do not look at only one side of a question, and that the one which pleases you? That you do not take up your opinions at second hand, from some book or some newspaper, which after all only reflects your own feelings, your own opinions? You should ask yourselves that question, seriously and often: "Are my thoughts really free?" No one values more highly than I do the advantage of a free press. But you must remember always that a newspaper editor, however honest or able, is no more infallible than the Pope; that he may, just as you may, only see one side of a question, while any question is sure to have two sides, or perhaps three or four; and if you only see the side which suits you, day after day, month after month, you must needs become bigoted to it. Your thoughts must needs run in one groove. They cannot (as Mr. Matthew Arnold would say) "play freely round" a question; and look it all over, boldly, patiently, rationally, charitably.

And I tell you that if you, or I, or any man, want to let our thoughts play freely round questions, and so escape from the tendency to become bigoted and narrow-minded which there is in every human being, then we must acquire something of that inductive habit of mind which the study of Natural Science gives. It is, after all, as Professor Huxley says, only common sense well regulated. But then it is well regulated; and how precious it is, if you can but get it. The art of seeing, the art of knowing what you see; the art of comparing, of perceiving true likenesses and true differences, and so of classifying and arranging what you see: the art of connecting facts together in your own mind in chains of cause and effect, and that accurately, patiently, calmly, without prejudice, vanity, or temper— this is what is wanted for true freedom of mind. But accuracy, patience, freedom from prejudice, carelessness for all except the truth, whatever the truth may be—are not these the virtues of a truly free spirit? Then, as I said just now, I know no study so able to give that free habit of mind as the Study of Natural Science.

Equality, too: whatever equality may or may not be just, or possible; this at least, is just, and I hope possible; that every man, every child, of every rank, should have an equal chance of education; an equal chance of developing all that is in him by nature; an equal chance of acquiring a fair knowledge of those facts of the universe which specially concern him; and of having his reason trained to judge of them. I say, whatever equal rights men may or may not have, they have this right. Let every boy, every girl, have an equal and sound education. If I had my way, I would give the same education to the child of the collier and to the child of a peer. I would see that they were taught the same things, and by the same method. Let them all begin alike, say I. They will be handicapped heavily enough as they go on in life, without our handicapping them in their first race. Whatever stable they come out of, whatever promise they show, let them all train alike, and start fair, and let the best colt win.

Well: but there is a branch of education in which, even now, the poor man can compete fairly against the rich; and that is, Natural Science. In the first place, the rich, blind to their own interest, have neglected it hitherto in their schools; so that they have not the start of the poor man on that subject which they have on many. In the next place, Natural Science is a subject which a man cannot learn by paying for teachers. He must teach it himself, by patient observation, by patient common sense. And if the poor man is not the rich man's equal in those qualities, it must be his own fault, not his purse's. Many shops have I seen about the world, in which fools could buy articles more or less helpful to them; but never saw I yet an observation-shop, nor a common-sense shop either. And if any man says, "We must buy books:" I answer, a poor man now can obtain better scientific books than a duke or a prince could sixty years ago, simply because then the books did not exist. When I was a boy I would have given much, or rather my father would have given much, if I could have got hold of such scientific books as are to be found now in any first-class elementary school. And if more expensive books are needed; if a microscope or apparatus is needed; can you not get them by the co-operative method, which has worked so well in other matters? Can you not form yourselves into a Natural Science club, for buying such things and lending them round among your members; and for discussion also, the reading of scientific papers of your own writing, the comparing of your observations, general mutual help and mutual instructions? Such societies are becoming numerous now, and gladly should I see one in every town. For in science, as in most matters, "As iron sharpeneth iron, so a man sharpeneth the countenance of his friend."

And Brotherhood: well, if you want that; if you want to mix with men, and men, too, eminently worth mixing with, on the simple ground that "a man's a man for a' that;" if you want to become the acquaintances, and—if you prove worthy—the friends, of men who will be glad to teach you all they know, and equally glad to learn from you anything you can teach them, asking no questions about you, save, first—Is he an honest student of Nature for her own sake? And next—Is he a man who will not quarrel, or otherwise behave in an unbrotherly fashion to his fellow-students?—If you want a ground of brotherhood with men, not merely in these islands, but in America, on the Continent—in a word, all over the world—such as rank, wealth, fashion, or other artificial arrangements of the world cannot give and cannot take away; if you want to feel yourself as good as any man in theory, because you are as good as any man in practice, except those who are better than you in the same line, which is open to any and every man; if you wish to have the inspiring and ennobling feeling of being a brother in a great freemasonry which owns no difference of rank, of creed, or of nationality—the only freemasonry, the only International League which is likely to make mankind (as we all hope they will be some day) one—then become men of science. Join the freemasonry in which Hugh Miller, the poor Cromarty stonemason, in which Michael Faraday, the poor bookbinder's boy, became the companions and friends of the noblest and most learned on earth, looked up to by them not as equals merely but as teachers and guides, because philosophers and discoverers.

Do you wish to be great? Then be great with true greatness; which is,—knowing the facts of nature, and being able to use them. Do you wish to be strong? Then be strong with true strength; which is, knowing the facts of nature, and being able to use them. Do you wish to be wise? Then be wise with true wisdom; which is, knowing the facts of nature, and being able to use them. Do you wish to be free? Then be free with true freedom; which is again, knowing the facts of nature, and being able to use them.

I dare say some of my readers, especially the younger ones, will demur to that last speech of mine. Well, I hope they will not be angry with me for saying it. I, at least, shall certainly not he angry with them. For when I was young I was very much of what I suspect is their opinion. I used to think one could get perfect freedom, and social reform, and all that I wanted, by altering the arrangements of society and legislation; by constitutions, and Acts of Parliament; by putting society into some sort of freedom-mill, and grinding it all down, and regenerating it so. And that something can be done by improved arrangements, something can be done by Acts of Parliament, I hold still, as every rational man must hold.

But as I grew older, I began to see that if things were to be got right, the freedom-mill would do very little towards grinding them right, however well and amazingly it was made. I began to see that what sort of flour came out at one end of the mill, depended mainly on what sort of grain you had put in at the other; and I began to see that the problem was to get good grain, and then good flour would be turned out, even by a very clumsy old-fashioned sort of mill. And what do I mean by good grain? Good men, honest men, accurate men, righteous men, patient men, self-restraining men, fair men, modest men. Men who are aware of their own vast ignorance compared with the vast amount that there is to be learned in such a universe as this. Men who are accustomed to look at both sides of a question; who, instead of making up their minds in haste like bigots and fanatics, wait like wise men, for more facts, and more thought about the facts. In one word, men who had acquired just the habit of mind which the study of Natural Science can give, and must give; for without it there is no use studying Natural Science; and the man who has not got that habit of mind, if he meddles with science, will merely become a quack and a charlatan, only fit to get his bread as a spirit-rapper, or an inventor of infallible pills.

And when I saw that, I said to myself—I will train myself, by Natural Science, to the truly rational, and therefore truly able and useful, habit of mind; and more, I will, for it is my duty as an Englishman, train every Englishman over whom I can get influence in the same scientific habit of mind, that I may, if possible, make him, too, a rational and an able man.

And, therefore, knowing that most of you, my readers—probably all of you, as you ought and must if you are Britons, think much of social and political questions—therefore, I say, I entreat you to cultivate the scientific spirit by which alone you can judge justly of those questions. I ask you to learn how to "conquer nature by obeying her," as the great Lord Bacon said two hundred and fifty years ago. For so only will you in your theories and your movements, draw "bills which nature will honour"—to use Mr. Carlyle's famous parable—because they are according to her unchanging laws, and not have them returned on your hands, as too many theorists' are, with "no effects" written across their backs.

Take my advice for yourselves, dear readers, and for your children after you; for, believe me, I am showing you the way to true and useful, and, therefore, to just and deserved power. I am showing you the way to become members of what I trust will be—what I am certain ought to be—the aristocracy of the future.

I say it deliberately, as a student of society and of history. Power will pass more and more, if all goes healthily and well, into the hands of scientific men; into the hands of those who have made due use of that great heirloom which the philosophers of the seventeenth century left for the use of future generations, and specially of the Teutonic race.

For the rest, events seem but too likely to repeat themselves again and again all over the world, in the same hopeless circle. Aristocracies of mere birth decay and die, and give place to aristocracies of mere wealth; and they again to "aristocracies of genius," which are really aristocracies of the noisiest, of mere scribblers and spouters, such as France is writhing under at this moment. And when these last have blown off their steam, with mighty roar, but without moving the engine a single yard, then they are but too likely to give place to the worst of all aristocracies, the aristocracy of mere "order," which means organised brute force and military despotism. And, after that, what can come, save anarchy, and decay, and social death?

What else?—unless there be left in the nation, in the society, as the salt of the land, to keep it all from rotting, a sufficient number of wise men to form a true working aristocracy, an aristocracy of sound and rational science? If they be strong enough (and they are growing stronger day by day over the civilised world), on them will the future of that world mainly depend. They will rule, and they will act—cautiously we may hope, and modestly and charitably, because in learning true knowledge they will have learnt also their own ignorance, and the vastness, the complexity, the mystery of nature. But they will be able to rule, they will be able to act, because they have taken the trouble to learn the facts and the laws of nature. They will rule; and their rule, if they are true to themselves, will be one of health and wealth, and peace, of prudence and of justice. For they alone will be able to wield for the benefit of man the brute forces of nature; because they alone will have stooped, to "conquer nature by obeying her."

So runs my dream. I ask my young readers to help towards making that dream a fact, by becoming (as many of them as feel the justice of my words) honest and earnest students of Natural Science.

But now: why should I, as a clergyman, interest myself specially in the spread of Natural Science? Am I not going out of my proper sphere to meddle with secular matters? Am I not, indeed, going into a sphere out of which I had better keep myself, and all over whom I may have influence? For is not science antagonistic to religion? and, if so, what has a clergyman to do, save to warn the young against it, instead of attracting them towards it?

First, as to meddling with secular matters. I grudge that epithet of "secular" to any matter whatsoever. But I do more; I deny it to anything which God has made, even to the tiniest of insects, the most insignificant atom of dust. To those who be-

19

lieve in God, and try to see all things in God, the most minute
natural phenomenon cannot be secular. It must be divine; I
say, deliberately, divine; and I can use no less lofty word. The
grain of dust is a thought of God; God's power made it; God's
wisdom gave it whatsoever properties or qualities it may pos-
sess; God's providence has put it in the place where it is now,
and has ordained that it should be in that place at that moment, by a train of causes and effects which reaches back to the
very creation of the universe. The grain of dust can no more go
from God's presence, or flee from God's Spirit, than you or I
can. If it go up to the physical heaven, and float (as it actually
often does) far above the clouds, in those higher strata of the
atmosphere which the aeronaut has never visited, whither the
Alpine snow-peaks do not rise, even there it will be obeying
physical laws which we term hastily laws of Nature, but which
are really the laws of God: and if it go down into the physical
abyss; if it be buried fathoms, miles, below the surface, and become an atom of some rock still in the process of consolidation,
has it escaped from God, even in the bowels of the earth? Is
it not there still obeying physical laws, of pressure, heat, crystallisation, and so forth, which are laws of God—the will and
mind of God concerning particles of matter? Only look at all
created things in this light—look at them as what they are, the
expressions of God's mind and will concerning this universe
in which we live—"the Word of God," as Bacon says, "revealed
in facts"—and then you will not fear physical science; for you
will be sure that, the more you know of physical science, the
more you will know of the works and of the will of God. At
least, you will be in harmony with the teaching of the Psalmist: "The heavens," says he, "declare the glory of God; and the
firmament showeth His handiwork. There is neither speech
nor language where their voices are not heard among them."
So held the Psalmist concerning astronomy, the knowledge of
the heavenly bodies; and what he says of sun and stars is true
likewise of the flowers around our feet, of which the greatest
Christian poet of modern times has said—

> *To me the meanest flower that grows may give*
> *Thoughts that do lie too deep for tears.*

So, again, you will be in harmony with the teaching of St. Paul, who told the Romans "that the invisible things of God are clearly seen from the creation of the-world, being understood by the things that are made, even His eternal power and God-head;" and who told the savages of Lycaonia that "God had not left Himself without witness, in that He did good and sent men rain from heaven, and fruitful seasons, filling men's hearts with food and gladness." Rain and fruitful seasons witnessed to all men of a Father in heaven. And he who wishes to know how truly St. Paul spoke, let him study the laws which produce and regulate rain and fruitful seasons, what we now call climatology, meteorology, geography of land and water. Let him read that truly noble Christian work, Maury's "Physical Geography of the Sea;" and see, if he be a truly rational man, how advanced science, instead of disproving, has only corroborated St. Paul's assertion, and how the ocean and the rain-cloud, like the sun and stars, declare the glory of God. And if anyone undervalues the sciences which teach us concerning stones and plants and animals, or thinks that nothing can be learnt from them concerning God—allow one who has been from childhood only a humble, though he trusts a diligent student of these sciences—allow him, I say, to ask in all reverence, but in all frankness, who it was who said, "Consider the lilies of the field, how they grow." "Consider the birds of the air—and how your Heavenly Father feedeth them."

Consider them. If He has bid you do so, can you do so too much?

I know, of course, the special application which our Lord made of these words. But I know, too, from experience, that the more you study nature, in all her forms the more you will find that the special application itself is deeper, wider, more literally true, more wonderful, more tender, and if I dare use such a word, more poetic, than the unscientific man can guess.

But let me ask you further—do you think that our Lord in that instance, and in those many instances in which He drew his parables and lessons from natural objects, was leading men's minds on to dangerous ground, and pointing out to them a subject of contemplation in the laws and processes of the natural world, and their analogy with those of the spiritual world, the kingdom of God—a subject of contemplation, I say, which it was not safe to contemplate too much?

I appeal to your common sense. If He who spoke these words were (as I believe) none other than the Creator of the universe, by whom all things were made, and without whom nothing was made that is made, do you suppose that He would have bid you to consider His universe, had it been dangerous for you to do so?

Do you suppose, moreover, that the universe, which He, the Truth, the Light, the Love, has made, can be otherwise then infinitely worthy to be considered? or that the careful, accurate, and patient consideration of it, even to its minutest details, can be otherwise than useful to man, and can bear witness of aught, save the mind and character of Him who made it? And if so, can it be a work unfit for, unworthy of, a clergyman—whose duty is to preach Him to all, and in all ways,—to call on men to consider that physical world which, like the spiritual world, consists, holds together, by Him, and lives and moves and has its being in Him?

And here I must pause to answer an objection which I have heard in my youth from many pious and virtuous people—better people in God's sight, than I, I fear, can pretend to be.

They used to say, "This would be all very true if there were not a curse upon the earth." And then they seemed to deduce, from the fact of that curse, a vague notion (for it was little more) that this world was the devil's world, and that therefore physical facts could not be trusted, because they were disordered, and deceptive, and what not.

Now, in justice to the Bible, and in justice to the Church of England, I am bound to say that such a statement, or anything like it, is contrary to the doctrines of both. It is contrary to Scripture. According to it, the earth is not cursed. For it is said in Gen. viii. 21, "And the Lord said, I will not again curse the ground any more for man's sake. While the earth remaineth, seed-time and harvest, cold and heat, summer and winter, day and night shall not cease." According to Scripture, again, physical facts are not disordered. The Psalmist says, "They continue this day according to their ordinance; for all things serve Thee." And again, "Thou hast made them fast for ever and ever. Thou hast given them a law which cannot be broken."

So does the Bible (not to quote over again the passages which I have already given you from St. Paul, and One greater than St. Paul) declare the permanence of natural laws, and the trustworthiness of natural phenomena as obedient to God. And so does the Church of England. For she has incorporated into her services that magnificent hymn, which our forefathers called the Song of the Three Children; which is, as it were, the very flower and crown of the Old Testament; the summing up of all that is true and eternal in the old Jewish faith; as true for us as for them: as true millions of years hence as it is now—which cries to all heaven and earth, from the skies above our heads to the green herb beneath our feet, "O all ye works of the Lord, bless ye the Lord; praise Him and magnify Him for ever." On that one hymn I take my stand. That is my charter as a student of Natural Science. As long as that is sung in an English church, I have a right to investigate Nature boldly without stint or stay, and to call on all who have the will, to investigate her boldly likewise, and with Socrates of old, to follow the Logos whithersoever it leads.

The Logos. I must pause on that word. It meant at first, no doubt, simply speech, argument, reason. In the mind of Socrates it had a deeper meaning, at which he only dimly guessed; which was seen more clearly by Philo and the Alexandrian Jews; which was revealed in all its fulness to the beloved Apostle St. John, till he gathered speech to tell men of a Logos, a Word, who was in the beginning with God, and was God; by whom all things were made, and without Him was not anything made that was made; and how in Him was Life, and the Life was the light of men; and that He was none other than Jesus Christ our Lord.

Yes, that is the truth. And to that truth no man can add, and from it no man can take away. And as long as we believe that as long as we believe that in His light alone can we see light— as long as we believe that the light around us, whether physical or spiritual, is given by Him without whom nothing is made— so long we shall not fear to meet Light, so long we shall not fear to investigate Life; for we shall know, however strange or novel, beautiful or awful, the discoveries we make may be, we are only following the Word whithersoever He may lead us; and that He can never lead us amiss

# I. THE SOIL OF THE FIELD[2]

My dear readers, let me, before touching on the special subject of this paper, say a few words on that of the whole series.

It is geology: that is, the science which explains to us the *rind* of the earth; of what it is made; how it has been made. It tells us nothing of the mass of the earth. That is, properly speaking, an astronomical question. If I may be allowed to liken this earth to a fruit, then astronomy will tell us—when it knows—how the fruit grew, and what is inside the fruit. Geology can only tell us at most how its rind, its outer covering, grew, and of what it is composed; a very small part, doubtless, of all that is to be known about this planet.

But as it happens, the mere rind of this earth-fruit which has, countless ages since, dropped, as it were, from the Bosom of God, the Eternal Fount of Life—the mere rind of this earth-fruit, I say, is so beautiful and so complex, that it is well worth our awful and reverent study. It has been well said, indeed, that the history of it, which we call geology, would be a magnificent epic poem, were there only any human interest in it; did it deal with creatures more like ourselves than stones, and bones, and the dead relics of plants and beasts. Whether there be no human interest in geology; whether man did not exist

2    These Lectures were delivered to the members of the Natural Science Class at Chester in 1871.

on the earth during ages which have seen enormous geological changes, is becoming more and more an open question.

But meanwhile all must agree that there is matter enough for interest—nay, room enough for the free use of the imagination, in a science which tells of the growth and decay of whole mountain-ranges, continents, oceans, whole tribes and worlds of plants and animals.

And yet it is not so much for the vastness and grandeur of those scenes of the distant past, to which the science of geology introduces us, that I value it as a study, and wish earnestly to awaken you to its beauty and importance. It is because it is the science from which you will learn most easily a sound scientific habit of thought. I say most easily; and for these reasons. The most important facts of geology do not require, to discover them, any knowledge of mathematics or of chemical analysis; they may be studied in every bank, every grot, every quarry, every railway-cutting, by anyone who has eyes and common sense, and who chooses to copy the late illustrious Hugh Miller, who made himself a great geologist out of a poor stonemason. Next, its most important theories are not, or need not be, wrapped up in obscure Latin and Greek terms. They may be expressed in the simplest English, because they are discovered by simple common sense. And thus geology is (or ought to be), in popular parlance, the people's science—the science by studying which, the man ignorant of Latin, Greek, mathematics, scientific chemistry, can yet become—as far as his brain enables him—a truly scientific man.

But how shall we learn science by mere common sense?

First. Always try to explain the unknown by the known. If you meet something which you have not seen before, then think of the thing most like it which you have seen before; and try if that which you know explains the one will not explain the other also. Sometimes it will; sometimes it will not. But if it will, no one has a right to ask you to try any other explanation.

Suppose, for instance, that you found a dead bird on the top of a cathedral tower, and were asked how you thought it had got there. You would say, "Of course, it died up here." But if a friend said, "Not so; it dropped from a balloon, or from the clouds;" and told you the prettiest tale of how the bird came to so strange an end, you would answer, "No, no; I must reason from what I know. I know that birds haunt the cathedral tower; I know that birds die; and therefore, let your story be as pretty as it may, my common sense bids me take the simplest explanation, and say—it died here." In saying that, you would be talking scientifically. You would have made a fair and sufficient induction (as it is called) from the facts about birds' habits and birds' deaths which you know.

But suppose that when you took the bird up you found that it was neither a jackdaw, nor a sparrow nor a swallow, as you expected, but a humming-bird. Then you would be adrift again. The fact of it being a humming-bird would be a new fact which you had not taken into account, and for which your old explanation was not sufficient; and you would have to try a new induction—to use your common sense afresh—saying, "I have not to explain merely how a dead bird got here, but how a dead humming-bird."

And now, if your imaginative friend chimed in triumphantly with: "Do you not see that I was right after all? Do you not see that it fell from the clouds? that it was swept away hither, all the way from South America, by some south-westerly storm, and wearied out at last, dropped here to find rest, as in a sacred-place?" what would you answer? "My friend, that is a beautiful imagination; but I must treat it only as such, as long as I can explain the mystery more simply by facts which I do know. I do not know that humming-birds can be blown across the Atlantic alive. I do know they are actually brought across the Atlantic dead; are stuck in ladies' hats. I know that ladies visit the cathedral; and odd as the accident is, I prefer to believe, till I get a better explanation, that the humming-bird has simply dropped out of a lady's hat." There, again, you would be speaking common sense; and using, too, sound inductive method; trying to explain what you do not know from what you do know already.

Now, I ask of you to employ the same common sense when you read and think of Geology.

It is very necessary to do so. For in past times men have tried to explain the making of the world around them, its oceans, rivers, mountains, and continents, by I know not what of fancied cataclysms and convulsions of nature; explaining the unknown by the still more unknown, till some of their geological theories were no more rational, because no more founded on known facts, than that of the New Zealand Maories, who hold that some god, when fishing, fished up their islands out of the bottom of the ocean. But a sounder and wiser school of geologists now reigns; the father of whom, in England at least, is the venerable Sir Charles Lyell. He was almost the first of Englishmen who taught us to see—what common sense tells us—that the laws which we see at work around us now have been most probably at work since the creation of the world; and that whatever changes may seem to have taken place in past ages, and in ancient rocks, should be explained, if possible, by the changes which are taking place now in the most recent deposits—in the soil of the field.

And in the last forty years—since that great and sound idea has become rooted in the minds of students, and especially of English students, geology has thriven and developed, perhaps more than any other science; and has led men on to discoveries far more really astonishing and awful than all fancied convulsions and cataclysms.

I have planned this series of papers, therefore, on Sir Charles Lyell's method. I have begun by trying to teach a little about the part of the earth's crust which lies nearest us, which we see most often; namely, the soil; intending, if my readers do me the honour to read the papers which follow, to lead them downward, as it were, into the earth; deeper and deeper in each paper, to rocks and minerals which are probably less known to them than the soil in the fields. Thus you will find I shall lead you, or try to lead you on, throughout the series, from the known to the unknown, and show you how to explain the latter by the former. Sir Charles Lyell has, I see, in the new edition of his "Student's Elements of Geology," begun his book with the uppermost, that is, newest, strata, or layers; and has gone regularly downwards in the course of the book to the lowest or earliest strata; and I shall follow his plan.

I must ask you meanwhile to remember one law or rule, which seems to me founded on common sense; namely, that the uppermost strata are really almost always the newest; that when two or more layers, whether of rock or earth—or indeed two stones in the street, or two sheets on a bed, or two books on a table—any two or more lifeless things, in fact, lie one on the other, then the lower one was most probably put there first, and the upper one laid down on the lower. Does that seem to you a truism? Do I seem almost impertinent in asking you to remember it? So much the better. I shall be saved unnecessary trouble hereafter.

But some one may say, and will have a right to say, "Stop—the lower thing may have been thrust under the upper one." Quite true: and therefore I said only that the lower one was most probably put there first. And I said "most probably," because it is most probable that in nature we should find things done by the method which costs least force, just as you do them. I will warrant that when you want to hide a thing, you lay something down on it ten times for once that you thrust it under something else. You may say, "What? When I want to hide a paper, say, under the sofa-cover, do I not thrust it under?"

No, you lift up the cover, and slip the paper in, and let the cover fall on it again. And so, even in that case, the paper has got into its place first.

Now why is this? Simply because in laying one thing on another you only move weight. In thrusting one thing under another, you have not only to move weight, but to overcome friction. That is why you do it, though you are hardly aware of it: simply because so you employ less force, and take less trouble.

And so do clays and sands and stones. They are laid down on each other, and not thrust under each other, because thus less force is expended in getting them into place.

There are exceptions. There are cases in which nature does try to thrust one rock under another. But to do that she requires a force so enormous, compared with what is employed in laying one rock on another, that (so to speak) she continually fails; and instead of producing a volcanic eruption, produces only an earthquake. Of that I may speak hereafter, and may tell you, in good time, how to distinguish rocks which have been thrust in from beneath, from rocks which have been laid down from above, as every rock between London and Birmingham or Exeter has been laid down. That I only assert now. But I do not wish you to take it on trust from me. I wish to prove it to you as I go on, or to do what is far better for you: to put you in the way of proving it for yourself, by using your common sense.

At the risk of seeming prolix, I must say a few more words on this matter. I have special reasons for it. Until I can get you to "let your thoughts play freely" round this question of the superposition of soils and rocks, there will be no use in my going on with these papers.

Suppose then (to argue from the known to the unknown) that you were watching men cleaning out a pond. Atop, perhaps, they would come to a layer of soft mud, and under that to a layer of sand. Would not common sense tell you that the sand was there first, and that the water had laid down the mud on the top of it? Then, perhaps, they might come to a layer of dead leaves. Would not common sense tell you that the leaves were there before the sand above them? Then, perhaps, to a layer of mud again. Would not common sense tell you that the mud was there before the leaves? And so on down to the bottom of the pond, where, lastly, I think common sense would tell you that the bottom of the pond was there already, before all the layers which were laid down on it. Is not that simple common sense?

Then apply that reasoning to the soils and rocks in any spot on earth. If you made a deep boring, and found, as you would in many parts of this kingdom, that the boring, after passing through the soil of the field, entered clays or loose sands, you would say the clays were there before the soil. If it then went down into sandstone, you would say—would you not?—that sandstone must have been here before the clay; and however thick—even thousands of feet—it might be, that would make no difference to your judgment. If next the boring came into quite different rocks; into a different sort of sandstone and shales, and among them beds of coal, would you not say— These coal-beds must have been here before the sandstones? And if you found in those coal-beds dead leaves and stems of plants, would you not say—Those plants must have been laid down here before the layers above them, just as the dead leaves in the pond were?

If you then came to a layer of limestone, would you not say the same? And if you found that limestone full of shells and corals, dead, but many of them quite perfect, some of the corals plainly in the very place in which they grew, would you not say—These creatures must have lived down here before the coal was laid on top of them? And if, lastly, below the limestone you came to a bottom rock quite different again, would you not say—The bottom rock must have been here before the rocks on the top of it?

And if that bottom rock rose up a few miles off, two thousand feet, or any other height, into hills, what would you say then? Would you say: "Oh, but the rock is not bottom rock; is not under the limestone here, but higher than it. So perhaps in this part it has made a shift, and the highlands are younger than the lowlands; for see, they rise so much higher?" Would not that be as wise as to say that the bottom of the pond was not there before the pond mud, because the banks round the pond rose higher than the mud?

Now for the soil of the field.

If we can understand a little about it, what it is made of, and how it got there, we shall perhaps be on the right road toward understanding what all England—and, indeed, the crust of this whole planet—is made of; and how its rocks and soils got there.

But we shall best understand how the soil in the field was made, by reasoning, as I have said, from the known to the unknown. What do I mean? This: On the uplands are fields in which the soil is already made. You do not know how? Then look for a field in which the soil is still being made. There are plenty in every lowland. Learn how it is being made there; apply the knowledge which you learn from them to the upland fields which are already made.

If there is, as there usually is, a river-meadow, or still better, an æstuary, near your town, you have every advantage for seeing soil made. Thousands of square feet of fresh-made soil spread between your town and the sea; thousands more are in process of being made.

You will see now why I have begun with the soil in the field; because it is the uppermost, and therefore latest, of all the layers; and also for this reason, that, if Sir Charles Lyell's theory be true—as it is—then the soils and rocks below the soil of the field may have been made in the very same way in which the soil of the field is made. If so, it is well worth our while to examine it.

You all know from whence the soil comes which has filled up, in the course of ages, the great æstuaries below London, Stirling, Chester, or Cambridge.

It is river mud and sand. The river, helped by tributary brooks right and left, has brought down from the inland that enormous mass. You know that. You know that every flood and freshet brings a fresh load, either of fine mud or of fine sand, or possibly some of it peaty matter out of distant hills. Here is one indisputable fact from which to start. Let us look for another.

How does the mud get into the river? The rain carries it thither.

If you wish to learn the first elements of geology by direct experiment, do this: The next rainy day—the harder it rains the better—instead of sitting at home over the fire, and reading a book about geology, put on a macintosh and thick boots, and get away, I care not whither, provided you can find there running water. If you have not time to get away to a hilly country,

then go to the nearest bit of turnpike road, or the nearest sloping field, and see in little how whole continents are made, and
unmade again. Watch the rain raking and sifting with its million delicate fingers, separating the finer particles from the
coarser, dropping the latter as soon as it can, and carrying the
former downward with it toward the sea. Follow the nearest
roadside drain where it runs into a pond, and see how it drops
the pebbles the moment it enters the pond, and then the sand
in a fan-shaped heap at the nearest end; but carries the fine
mud on, and holds it suspended, to be gradually deposited
at the bottom in the still water; and say to yourself: Perhaps
the sands which cover so many inland tracts were dropped by
water, very near the shore of a lake or sea, and by rapid currents. Perhaps, again, the brick clays, which are often mingled
with these sands, were dropped, like the mud in the pond, in
deeper water farther from the shore, and certainly in stilt
water. But more. Suppose once more, then, that looking and
watching a pond being cleared out, under the lowest layer of
mud, you found—as you would find in any of those magnificent reservoirs so common in the Lancashire hills—a layer of
vegetable soil, with grass and brushwood rooted in it. What
would you say but: The pond has not been always full. It has
at some time or other been dry enough to let a whole copse
grow up inside it?

And if you found—as you will actually find along some English
shores—under the sand hills, perhaps a bed of earth with shells
and bones; under that a bed of peat; under that one of blue silt;
under that a buried forest, with the trees upright and rooted;
under that another layer of blue silt full of roots and vegetable
fibre; perhaps under that again another old land surface with
trees again growing in it; and under all the main bottom clay
of the district—what would common sense tell you? I leave you
to discover for yourselves. It certainly would not tell you that
those trees were thrust in there by a violent convulsion, or that
all those layers were deposited there in a few days, or even a
few years; and you might safely indulge in speculations about
the antiquity of the æstuary, and the changes which it has un-
dergone, with which I will not frighten you at present.

It will be fair reasoning to argue thus. You may not be always right in your conclusion, but still you will be trying fairly to explain the unknown by the known.

But have Rain and Rivers alone made the soil?

How very much they have done toward making it you will be able to judge for yourselves, if you will read the sixth chapter of Sir Charles Lyell's new "Elements of Geology," or the first hundred pages of that admirable book, De la Beche's "Geological Observer;" and last, but not least, a very clever little book called "Rain and Rivers," by Colonel George Greenwood.

But though rain, like rivers, is a carrier of soil, it is more. It is a maker of soil, likewise; and by it mainly the soil of an upland field is made, whether it be carried down to the sea or not.

If you will look into any quarry you will see that however compact the rock may be a few feet below the surface, it becomes, in almost every case, rotten and broken up as it nears the upper soil, till you often cannot tell where the rock ends and the soil begins.

Now this change has been produced by rain. First, mechanically, by rain in the shape of ice. The winter rain gets into the ground, and does by the rock what it has done by the stones of many an old building. It sinks into the porous stone, freezes there, expands in freezing, and splits and peels the stone with a force which is slowly but surely crumbling the whole of Northern Europe and America to powder.

Do you doubt me? I say nothing but what you can judge of yourselves. The next time you go up any mountain, look at the loose broken stones with which the top is coated, just underneath the turf. What has broken them up but frost? Look again, as stronger proof, at the talus of broken stones—screes, as they call them in Scotland; rattles, as we call them in Devon—which lie along the base of many mountain cliffs. What has brought them down but frost? If you ask the country folk they will tell you whether I am right or not. If you go thither, not in the summer, but just after the winter's frost, you will see for yourselves, by the fresh frost-crop of newly-broken bits, that I am right. Possibly you may find me to be even more right than is desirable, by having a few angular stones, from the size of your head to that of your body, hurled at you by the frost-giants up above. If you go to the Alps at certain seasons, and hear the thunder of the falling rocks, and see their long lines—moraines, as they are called—sliding slowly down upon the surface of the glacier, then you will be ready to believe the geologist who tells you that frost, and probably frost alone, has hewn out such a peak as the Matterhorn from some vast table-land; and is hewing it down still, winter after winter, till some day, where the snow Alps now stand, there shall be rolling uplands of rich cultivable soil.

So much for the mechanical action of rain, in the shape of ice. Now a few words on its chemical action.

Rain water is seldom pure. It carries in it carbonic acid; and that acid, beating in shower after shower against the face of a cliff—especially if it be a limestone cliff—weathers the rock chemically; changing (in case of limestone) the insoluble carbonate of lime into a soluble bicarbonate, and carrying that away in water, which, however clear, is still hard. Hard water is usually water which has invisible lime in it; there are from ten to fifteen grains and more of lime in every gallon of limestone water. I leave you to calculate the enormous weight of lime which must be so carried down to the sea every year by a single limestone or chalk brook. You can calculate it, if you like, by ascertaining the weight of lime in each gallon, and the average quantity of water which comes down the stream in a day; and when your sum is done, you will be astonished to find it one not of many pounds, but probably of many tons, of solid lime, which you never suspected or missed from the hills around. Again, by the time the rain has sunk through the soil, it is still less pure. It carries with it not only carbonic acid, but acids produced by decaying vegetables—by the roots of the grasses and trees which grow above; and they dissolve the cement of the rock by chemical action, especially if the cement be lime or iron. You may see this for yourselves, again and again. You may see how the root of a tree, penetrating the earth, discolours the soil with which it is in contact. You may see how the whole rock, just below the soil, has often changed in colour from the compact rock below, if the soil be covered with a dense layer of peat or growing vegetables.

But there is another force at work, and quite as powerful as rain and rivers, making the soil of alluvial flats. Perhaps it has helped, likewise, to make the soil of all the lowlands in these isles—and that is, the waves of the sea.

If you ever go to Parkgate, in Cheshire, try if you cannot learn there a little geology.

Walk beyond the town. You find the shore protected for a long way by a sea-wall, lest it should be eaten away by the waves. What the force of those waves can be, even on that sheltered coast, you may judge—at least you could have judged this time last year—by the masses of masonry torn from their iron clampings during the gale of three winters since. Look steadily at those rolled blocks, those twisted stanchions, if they are there still; and then ask yourselves—it will be fair reasoning from the known to the unknown—What effect must such wave-power as that have had beating and breaking for thousands of years along the western coasts of England, Scotland, Ireland? It must have eaten up thousands of acres—whole shires, may be, ere now. Its teeth are strong enough, and it knows neither rest nor pity, the cruel hungry sea. Give it but time enough, and what would it not eat up? It would eat up, in the course of ages, all the dry land of this planet, were it not baffled by another counteracting force, of which I shall speak hereafter.

As you go on beyond the sea-wall, you find what it is eating up. The whole low cliff is going visibly. But whither is it going? To form new soil in the æstuary. Now you will not wonder how old harbours so often become silted up. The sea has washed the land into them. But more, the sea-currents do not allow the sands of the æstuary to escape freely out to sea. They pile it up in shifting sand-banks about the mouth of the æstuary. The prevailing sea-winds, from whatever quarter, catch up the sand, and roll it up into sand-hills. Those sand-hills are again eaten down by the sea, and mixed with the mud of the tide-flats, and so is formed a mingled soil, partly of clayey mud, partly of sand; such a soil as stretches over the greater part of all our lowlands.

Now, why should not that soil, whether in England or in Scotland, have been made by the same means as that of every æstuary.

You find over great tracts of East Scotland, Lancashire, Norfolk, etc., pure loose sand just beneath the surface, which looks as if it was blown sand from a beach. Is it not reasonable to suppose that it is? You find rising out of many lowlands, crags which look exactly like old sea-cliffs eaten by the waves, from the base of which the waters have gone back. Why should not those crags be old sea-cliffs? Why should we not, following our rule of explaining the unknown by the known, assume that such they are till someone gives us a sound proof that they are not; and say—These great plains of England and Scotland were probably once covered by a shallow sea, and their soils made as the soil of any tide-flat is being made now?

But you may say, and most reasonably "The tide-flats are just at the sea-level. The whole of the lowland is many feet above the sea; it must therefore have been raised out of the sea, according to your theory: and what proofs have you of that?"

Well, that is a question both grand and deep, on which I shall not enter yet; but meanwhile, to satisfy you that I wish to play fair with you, I ask you to believe nothing but what you can prove for yourselves. Let me ask you this: suppose that you had proof positive that I had fallen into the river in the morning; would not your meeting me in the evening be also proof positive that somehow or other I had in the course of the day got out of the river? I think you will accept that logic as sound.

Now if I can give you proof positive, proof which you can see with your own eyes, and handle with your own hands, and alas! often feel but too keenly with your own feet, that the whole of the lowlands were once beneath the sea; then will it not be

37

certain that, somehow or other, they must have been raised out of the sea again?

And that I propose to do in my next paper, when I speak of the pebbles in the street.

Meanwhile I wish you to face fairly the truly grand idea, which all I have said tends to prove true—that all the soil we see is made by the destruction of older soils, whether soft as clay, or hard as rock; that rain, rivers, and seas are perpetually melting and grinding up old land, to compose new land out of it; and that it must have been doing so, as long as rain, rivers, and seas have existed. "But how did the first land of all get made?" I can only reply: A natural question: but we can only answer that, by working from the known to the unknown. While we are finding out how these later lands were made and unmade, we may stumble on some hints as to how the first primeval continents rose out of the bosom of the sea.

And thus I end this paper. I trust it has not been intolerably dull. But I wanted at starting to show my readers something of the right way of finding out truth on this and perhaps on all subjects; to make some simple appeals to your common sense; and to get you to accept some plain rules founded on common sense, which will be of infinite use to both you and me in my future papers.

I hope, meanwhile, that you will agree with me, that there is plenty of geological matter to be seen and thought over in the neighbourhood of any town.

Be sure, that wherever there is a river, even a drain; and a stone quarry, or even a roadside bank; much more where there is a sea, or a tidal æstuary, there is geology enough to be learnt, to explain the greater part of the making of all the continents on the globe.

## II. THE PEBBLES IN THE STREET

If you, dear reader, dwell in any northern town, you will almost certainly see paving courts and alleys, and some-times—to the discomfort of your feet—whole streets, or set up as bournestones at corners, or laid in heaps to be broken up for road-metal, certain round pebbles, usually dark brown or speckled gray, and exceedingly tough and hard. Some of them will be very large—boulders of several feet in diameter. If you move from town to town, from the north of Scotland as far down as Essex on the east, or as far down as Shrewsbury and Wolverhampton (at least) on the west, you will still find these pebbles, but fewer and smaller as you go south. It matters not what the rocks and soils of the country round may be. How-ever much they may differ, these pebbles will be, on the whole, the same everywhere.

But if your town be south of the valley of the Thames, you will find, as far as I am aware, no such pebbles there. The gravels round you will be made up entirely of rolled chalk flints, and bits of beds immediately above or below the chalk. The blocks of "Sarsden" sandstone—those of which Stonehenge is built—and the "plum-pudding stones" which are sometimes found with them, have no kindred with the northern pebbles. They belong to beds above the chalk.

Now if, seeing such pebbles about your town, you inquire, like a sensible person who wishes to understand something of the spot on which he lives, whence they come, you will be shown either a gravel-pit or a clay-pit. In the gravel the pebbles and boulders lie mixed with sand, as they do in the railway cutting just south of Shrewsbury; or in huge mounds of fine sweet earth, as they do in the gorge of the Tay about Dunkeld, and all the way up Strathmore, where they form long grassy mounds— *tomauns* as they call them in some parts of Scotland—*askers* as they call them in Ireland. These mounds, with their sweet fresh turf rising out of heather and bog, were tenanted—so Scottish children used to believe—by fairies. He that was lucky might hear inside them fairy music, and, the jingling of the fairy horses' trappings. But woe to him if he fell asleep upon the mound, for he would be spirited away into fairyland for seven years, which would seem to him but one day. A strange fancy; yet not so strange as the actual truth as to what these mounds are, and how they came into their places.

Or again, you might find that your town's pebbles and boulders came out of a pit of clay, in which they were stuck, without any order or bedding, like plums and raisins in a pudding. This clay goes usually by the name of boulder-clay. You would see such near any town in Cheshire and Lancashire; or along Leith shore, near Edinburgh; or, to give one more instance out of hundreds, along the coast at Scarborough. If you walk along the shore southward of that town, you will see, in the gullies of the cliff, great beds of sticky clay, stuffed full of bits of every rock between the Lake mountains and Scarborough, from rounded pebbles of most ancient rock down to great angular fragments of ironstone and coal. There, as elsewhere, the great majority of the pebbles have nothing to do with the rock on which the clay happens to lie, but have come, some of them, from places many miles away.

Now if we find spread over a low land pebbles composed of rocks which are only found in certain high lands, is it not an act of common sense to say—These pebbles have come from the highlands? And if the pebbles are rounded, while the rocks like them in the highlands always break off in angular shapes, is it not, again, an act of mere common sense to say—These pebbles were once angular, and have been rubbed round, either in getting hither or before they started hither?

Does all this seem to you mere truism, my dear reader? If so, I am sincerely glad to hear it. It was not so very long ago that such arguments would have been considered not only no truisms, but not even common sense.

But to return, let us take, as an example, a sample of these boulder clay pebbles from the neighbourhood of Liverpool and Birkenhead, made by Mr. De Rance, the government geological surveyor:

> Granite, greenstone, felspar porphyry, felstone, quartz rock (all igneous rocks, that is, either formed by, or altered by volcanic heat, and almost all found in the Lake mountains), 37 per cent.

> Silurian grits (the common stones of the Lake mountains deposited by water), 43 per cent.

> Ironstone, 1 per cent.

> Carboniferous limestone, 5 per cent.

> Permian or Triassic sandstones, *i.e.* rocks immediately round Liverpool, 12 per cent.

Now, does not this sample show, as far as human common sense can be depended on, that the great majority of these stones come from the Lake mountains, sixty or seventy miles north of Liverpool? I think your common sense will tell you that these pebbles are not mere concretions; that is, formed out of the substance of the clay after it was deposited. The least knowledge of mineralogy would prove that. But, even if you are no mineralogist, common sense will tell you, that if they were all concreted out of the same clay, it is most likely that they would be all of the same kind, and not of a dozen or more different kinds. Common sense will tell you, also, that if they were all concreted out of the same clay, it is a most extraordinary coincidence, indeed one too strange to be believed, if any less strange explanation can be found—that they should have taken the composition of different rocks which are found all together in one group of mountains to the northward. You will surely say—If this be granite, it has most probably come from a granite mountain; if this be grit, from a grit-stone mountain, and so on with the whole list. Why—are we to go out of our way to seek improbable explanations, when there is a probable one staring us in the face?

Next—and this is well worth your notice—if you will examine the pebbles carefully, especially the larger ones, you will find that they are not only more or less rounded, but often scratched; and often, too, in more than one direction, two or even three sets of scratches crossing each other; marked, as a cat marks an elder stem when she sharpens her claws upon it; and that these scratches have not been made by the quarrymen's tools, but are old marks which exist—as you may easily prove for yourself—while the stone is still lying in its bed of clay. Would it not be an act of mere common sense to say— These scratches have been made by the sharp points of other stones which have rubbed against the pebbles somewhere, and somewhen, with great force?

So far so good. The next question is—How did these stones get into the clay? If we can discover that, we may also discover how they wore rounded and scratched. We must find a theory which will answer our question; and one which, as Professor Huxley would say, "will go on all-fours," that is, will explain all the facts of the case, and not only a few of them.

What, then, brought the stones?

We cannot, I think, answer that question, as some have tried to answer it, by saying that they were brought by Noah's flood. For it is clear, that very violent currents of water would be needed to carry boulders, some of them weighing many tons, for many miles. Now Scripture says nothing of any such violent currents; and we have no right to put currents, or any other imagined facts, into Scripture out of our own heads, and then argue from them as if not we, but the text of Scripture had asserted their existence.

But still, they may have been rolled hither by water. That theory certainly would explain their being rounded; though not their being scratched. But it will not explain their being found in the clay.

Recollect what I said in my first paper: that water drops its pebbles and coarser particles first, while it carries the fine clayey mud onward in solution, and only drops it when the water becomes still. Now currents of such tremendous violence as to carry these boulder stones onward, would have carried the mud for many miles farther still; and we should find the boulders, not in clay, but lying loose together, probably on a hard rock bottom, scoured clean by the current. That is what we find in the beds of streams; that is just what we do not find in this case.

But the boulders may have been brought by a current, and then the water may have become still, and the clay settled quietly round them. What? Under them as well as over them? On that theory also we should find them only at the bottom of the clay. As it is, we find them scattered anywhere and everywhere through it, from top to bottom. So that theory will not do. Indeed, no theory will do which supposes them to have been brought by water alone.

Try yourself, dear reader, and make experiments, with running water, pebbles, and mud. If you try for seven years, I believe, you will never contrive to make your pebbles lie about in your mud, as they lie about in every pit in the boulder clay.

Well then, there we are at fault, it seems. We have no explanation drawn from known facts which will do—unless we are to suppose, which I don't think you will do, that stones, clay, and all were blown hither along the surface of the ground, by primeval hurricanes, ten times worse than those of the West Indies, which certainly will roll a cannon a few yards, but cannot, surely, roll a boulder stone a hundred miles.

Now, suppose that there was a force, an agent, known—luckily for you, not to you—but known too well to sailors and travellers; a force which is at work over the vast sheets of land at both the north and south poles; at work, too, on every high mountain range in the world, and therefore a very common natural force; and suppose that this force would explain all the facts, namely—

How the stones got here;

How they were scratched and rounded;

How they were imbedded in clay;

because it is notoriously, and before men's eyes now, carrying great stones hundreds of miles, and scratching and rounding them also; carrying vast deposits of mud, too, and mixing up mud and stones just as we see them in the brick-pits,—Would not our common sense have a right to try that explanation?— to suspect that this force, which we do not see at work in Britain now, may have been at work here ages since? That would at least be reasoning from the known to the unknown. What state of things, then, do we find among the highest mountains; and over whole countries which, though not lofty, lie far enough north or south to be permanently covered with ice?

We find, first, an ice-cap or ice-sheet, fed by the winter's snows, stretching over the higher land, and crawling downward and outward by its own weight, along the valleys, as glaciers.

We find underneath the glaciers, first a *moraine profonde,* consisting of the boulders and gravel, and earth, which the glacier has ground off the hillsides, and is carrying down with it.

These stones, of course, grind, scratch, and polish each other; and in like wise grind, scratch, and polish the rock over which they pass, under the enormous weight of the superincumbent ice.

We find also, issuing from under each glacier a stream, carrying the finest mud, the result of the grinding of the boulders against each other and the glacier.

We find, moreover, on the surface of the glaciers, *moraines supérieures*—long lines of stones and dirt which had fallen

44

from neighbouring cliffs, and are now travelling downward with the glaciers.

Their fate, if the glacier ends on land, is what was to be expected. The stones from above the glacier fall over the ice-cliff at its end, to mingle with those thrown out from underneath the glacier, and form huge banks of boulders, called terminal moraines, while the mud runs off, as all who have seen glaciers know, in a turbid torrent.

Their fate, again, is what was to be expected if the glacier ends, as it commonly does in Arctic regions, in the sea. The ice grows out to sea-ward for more than a mile sometimes, about one-eighth of it being above water, and seven-eighths below, so that an ice-cliff one hundred feet high may project into water eight hundred feet deep. At last, when it gets out of its depth, the buoyancy of the water breaks it off in icebergs, which float away, at the mercy of tides and currents, often grounding again in shallower water, and ploughing the sea-bottom as they drag along it. These bergs carry stones and dirt, often in large quantities; so that, whenever a berg melts or capsizes, it strews its burden confusedly about the sea-floor.

Meanwhile the fine mud which is flowing out from under the ice goes out to sea likewise, colouring the water far out, and then subsiding as a soft tenacious ooze, in which the stones brought out by the ice are imbedded. And this ooze—so those who have examined it assert—cannot be distinguished from the brick-clay, or fossiliferous boulder-clay, so common in the North. A very illustrious Scandinavian explorer, visiting Edinburgh, declared, as soon as he saw the sections of boulder-clay exhibited near that city, that this was the very substance which he saw forming in the Spitzbergen ice-fiords.[3]

I have put these facts as simply and baldly as I can, in order that the reader may look steadily at them, without having his

---

3    See a most charming paper on "The Physics of Arctic Ice," by Dr. Robert Brown of Campster, published in the *Quarterly Journal of the Geological Society*, June, 1870. This article is so remarkable, not only for its sound scientific matter, but for the vividness and poetic beauty of its descriptions, that I must express a hope that the learned author will some day enlarge it, and publish it in a separate form.

attention drawn off, or his fancy excited, by their real poetry and grandeur. Indeed, it would have been an impertinence to have done otherwise; for I have never seen a live glacier, by land or sea, though I have seen many a dead one. And the public has had the opportunity, lately, of reading so many delightful books about "peaks, passes, and glaciers," that I am bound to suppose that many of my readers know as much, or more, about them than I do.

But let us go a step farther; and, bearing in our minds what live glaciers are like, let us imagine what a dead glacier would be like; a glacier, that is, which had melted, and left nothing but its skeleton of stones and dirt.

We should find the faces of the rock scored and polished, generally in lines pointing down the valleys, or at least outward from the centre of the highlands, and polished and scored most in their upland or weather sides. We should find blocks of rock left behind, and perched about on other rocks of a different kind. We should find in the valleys the old moraines left as vast deposits of boulder and shingle, which would be in time sawn through and sorted over by the rivers. And if the sea-bottom outside were upheaved, and became dry land, we should find on it the remains of the mud from under the glacier, stuck full of stones and boulders iceberg-dropped. This mud would be often very irregularly bedded; for it would have been disturbed by the ploughing of the icebergs, and mixed here and there with dirt which had fallen from them. Moreover, as the sea became shallower and the mud-beds got awash one after the other, they would be torn about, re-sifted, and re-shaped by currents and by tides, and mixed with shore-sand ground out of shingle-beach, thus making confusion worse confounded. A few shells, of an Arctic or northern type, would be found in it here and there. Some would have lived near those later beaches, some in deeper water in the ancient ooze, wherever the iceberg had left it in peace long enough for sea-animals to colonise and breed in it. But the general appearance of the dried sea-bottom would be a dreary and lifeless waste of sands, gravels, loose boulders, and boulder-bearing clays; and wherever a boss of bare rock still stood up, it would be found ground down, and probably polished and scored by the ponderous icebergs which had lumbered over it in their passage out to sea.

In a word, it would look exactly as vast tracts of the English, Scotch, and Irish lowlands must have looked before returning vegetation coated their dreary sands and clays with a layer of brown vegetable soil.

Thus, and I believe thus only, can we explain the facts connected with these boulder pebbles. No agent known on earth can have stuck them in the clay, save ice, which is known to do so still elsewhere.

No known agent can have scratched them as they are scratched, save ice, which is known to do so still elsewhere.

No known agent—certainly not, in my opinion, the existing rivers—can have accumulated the vast beds of boulders which lie along the course of certain northern rivers; notably along the Dee about Aboyne—save ice bearing them slowly down from the distant summits of the Grampians.

No known agent, save ice, can have produced those rounded, and polished, and scored, and fluted *rochers moutonnés* "sheep-backed rocks"—so common in the Lake district; so common, too, in Snowdon, especially between the two lakes of Llanberis; common in Kerry; to be seen anywhere, as far as I have ascertained, around the Scotch Highlands, where the turf is cleared away from an unweathered surface of the rock, in the direction in which a glacier would have pressed against it had one been there. Where these polishings and scorings are found in narrow glens, it is, no doubt, an open question whether some of them may not be the work of water. But nothing but the action of ice can have produced what I have seen in land-locked and quiet fords in Kerry—ice-flutings in polished rocks below high-water mark, so large that I could lie down in one of them. Nothing but the action of ice could produce what may be seen in any of our mountains—whole sheets of rock ground down into rounded flats, irrespective of the lie of the beds, not in valleys, but on the brows and summits of mountains, often ending abruptly at the edge of some sudden cliff, where the true work of water, in the shape of rain and frost, is actually destroying the previous work of ice, and fulfilling the rule laid down (I think by Professor Geikie in his delightful book on Scotch scenery as influenced by its geology), that ice planes down into flats, while water saws out into crags and gullies; and that the rain and frost are even now restoring Scotch scenery to something of that ruggedness and picturesqueness which it must have lost when it lay, like Greenland, under the indiscriminating grinding of a heavy sheet of ice.

Lastly; no known agent, save ice, will explain those perched boulders, composed of ancient hard rocks, which may be seen in so many parts of these islands and of the Continent. No water power could have lifted those stones, and tossed them up high and dry on mountain ridges and promontories, upon rocks of a totally different kind. Some of my readers surely recollect Wordsworth's noble lines about these mysterious wanderers, of which he had seen many a one about his native hills:

> *As a huge stone is sometimes seen to lie*
> *Couched on the bald top of an eminence,*
> *Wonder to all who do the same espy*
> *By what means it could thither come, and whence;*
> *So that it seems a thing endued with sense:*
> *Like a sea-beast crawled forth, that on a shelf*
> *Of rock or sand reposeth, there to sun itself.*

Yes; but the next time you see such a stone, believe that the wonder has been solved, and found to be, like most wonders in Nature, more wonderful than we guessed it to be. It is not a sea-beast which has crawled forth, but an ice-beast which has been left behind; lifted up thither by the ice, as surely as the famous Pierre-à-bot, forty feet in diameter, and hundreds of boulders more, almost as large as cottages, have been carried by ice from the distant Alps right across the lake of Neufchatel, and stranded on the slopes of the Jura, nine hundred feet above the lake.[4]

Thus, I think, we have accounted for facts enough to make it probable that Britain was once covered partly by an ice-sheet,

---

4     See Lyell, "Antiquity of Man," p. 294 *et seq.*

as Greenland is now, and partly, perhaps, by an icy sea. But, to make assurance more sure, let us look for new facts, and try whether our ice-dream will account for them also. Let us investigate our case as a good medical man does, by "verifying his first induction."

He says: At the first glance, I can see symptoms *a, b, c.* It is therefore probable that my patient has got complaint A. But if he has he ought to have symptom d also. If I find that, my guess will be yet more probable. He ought also to have symptom e, and so forth; and as I find successively each of these symptoms which are proper to A, my first guess will become more and more probable, till it reaches practical certainty.

Now let us do the same, and say—If this strange dream be true, and the lowlands of the North were once under an icy sea, ought we not to find sea-shells in their sands and clays? Not abundantly, of course. We can understand that the sea-animals would be too rapidly covered up in mud, and too much disturbed by icebergs and boulders, to be very abundant. But still, some should surely be found here and there.

Doubtless; and if my northern-town readers will search the boulder-clay pits near them, they will most probably find a few shells, if not in the clay itself, yet in sand-beds mixed with them, and probably underlying them. And this is a notable fact, that the more species of shells they find, the more they will find—if they work out their names from any good book of conchology—of a northern type; of shells which notoriously, at this day, inhabit the colder seas.

It is impossible for me here to enter at length on a subject on which a whole literature has been already written. Those who wish to study it may find all that they need know, and more, in Lyell's "Student's Elements of Geology," and in chapter xii. of his "Antiquity of Man." They will find that if the evidence of scientific conchologists be worth anything, the period can be pointed out in the strata, though not of course in time, at which these seas began to grow colder, and southern and Mediterranean shells to disappear, their places being taken by shells of a temperate, and at last of an Arctic climate; which last have since retreated either toward their native North, or into cold water at great depths. From Essex across to Wales, from Wales to the æstuary of the Clyde, this fact has been verified again and again. And in the search for these shells, a fresh fact, and a most startling one, was discovered. They are to be found not only in the clay of the lowlands, but at considerable heights up the hills, showing that, at some time or other, these hills have been submerged beneath the sea.

Let me give one example, which any tourist into Wales may see for himself. Moel Tryfaen is a mountain over Carnarvon. Now perched on the side of that mountain, fourteen hundred feet above the present sea-level, is an ancient sea-beach, five-and-thirty feet thick, lying on great ice-scratched boulders, which again lie on the mountain slates. It was discovered by the late Mr. Trimmer, now, alas! lost to Geology. Out of that beach fifty-seven different species of shells have been taken; eleven of them are now exclusively Arctic, and not found in our seas; four of them are still common to the Arctic seas and to our own; and almost all the rest are northern shells.

Fourteen hundred feet above the present sea: and that, it must be understood, is not the greatest height at which such shells may be found hereafter. For, according to Professor Ramsay, drift of the same kind as that on Moel Tryfaen is found at a height of two thousand three hundred feet.

Now I ask my readers to use their common sense over this astounding fact—which, after all, is only one among hundreds; to let (as Mr. Matthew Arnold would well say) their "thought play freely" about it; and consider for themselves what those shells must mean. I say not may, but must, unless we are to believe in a "Deus quidam deceptor," in a God who puts shells upon mountain-sides only to befool honest human beings, and gives men intellects which are worthless for even the simplest work. Those shells must mean that that mountain, and therefore the mountains round it, must have been once fourteen hundred feet at least lower than they are now. That the sea in which they were sunk was far colder than now. That icebergs brought and dropped boulders round their flanks. That upon those boulders a sea-beach formed, and that dead shells were beaten into it from a sea-bottom close by. That, and no less, Moel Tryfaen must mean.

But it must mean, also, a length of time which has been well called "appalling." A length of time sufficient to let the mountain sink into the sea. Then length of time enough to enable those Arctic shells to crawl down from the northward, settle, and propagate themselves generation after generation; then length of time enough to uplift their dead remains, and the beach, and the boulders, and all Snowdonia, fourteen hundred feet into the air. And if anyone should object that the last upheaval may have been effected suddenly by a few tremendous earthquakes, we must answer—We have no proof of it. Earthquakes upheave lands now only by slight and intermittent upward pulses; nay, some lands we know to rise without any earthquake pulses, but by simple, slow, upward swelling of a few feet in a century; and we have no reason, and therefore no right, to suppose that Snowdonia was upheaved by any means or at any rate which we do not witness now; and therefore we are bound to allow, not only that there was a past "age of ice," but that that age was one of altogether enormous duration.

But meanwhile some of you, I presume, will be ready to cry— Stop! It may be our own weakness; but you are really going on too fast and too far for our small imaginations. Have you not played with us, as well as argued with us, till you have inveigled us step by step into a conclusion which we cannot and will not believe? That all this land should have been sunk beneath an icy sea? That Britain should have been as Greenland is now? We can't believe it, and we won't.

If you say so, like stout common-sense Britons, who have a wholesome dread of being taken in with fine words and wild speculations, I assure you I shall not laugh at you even in private. On the contrary, I shall say—what I am sure every scientific man will say—So much the better. That is the sort of audience which we want, if we are teaching natural science. We do not want haste, enthusiasm, gobe-moucherie, as the

French call it, which is agape to snap up any new and vast fancy, just because it is new and vast. We want our readers to be slow, suspicious, conservative, ready to "gib," as we say of a horse, and refuse the collar up a steep place, saying—I must stop and think. I don't like the look of the path ahead of me. It seems an ugly place to get up. I don't know this road, and I shall not hurry over it. I must go back a few steps, and make sure. I must see whether it is the right road; whether there are not other roads, a dozen of them perhaps, which would do as well and better than this.

This is the temper which finds out truth, slowly, but once and for all; and I shall be glad, not sorry, to see it in my readers.

And I am bound to say that it has been by that temper that this theory has been worked out, and the existence of this past age of ice, or glacial epoch, has been discovered, through many mistakes, many corrections, and many changes of opinion about details, for nearly forty years of hard work, by many men, in many lands.

As a very humble student of this subject, I may say that I have been looking these facts in the face earnestly enough for more than twenty years, and that I am about as certain that they can only be explained by ice, as I am that my having got home by rail can only be explained by steam.

But I think I know what startles you. It is the being asked to believe in such an enormous change in climate, and in the height of the land above the sea. Well—it is very astonishing, appalling—all but incredible, if we had not the facts to prove it. But of the facts there can be no doubt. There can be no doubt that the climate of this northern hemisphere has changed enormously more than once. There can be no doubt that the distribution of land and water, the shape and size of its continents and seas, have changed again and again. There can be no doubt that, for instance, long before the age of ice, the whole North of Europe was much warmer than it is now.

Take Greenland, for instance. Disco Island lies in Baffin's Bay, off the west coast of Greenland, in latitude 70°, far within the

Arctic circle. Now there certain strata of rock, older than the ice, have not been destroyed by the grinding of the ice-cap; and they are full of fossil plants. But of what kind of plants? Of the same families as now grow in the warmer parts of the United States. Even a tulip-tree has been found among them. Now how is this to be explained?

Either we must say that the climate of Greenland was then so much warmer than now, that it had summers probably as hot as those of New York; or we must say that these leaves and stems were floated thither from the United States. But if we say the latter, we must allow a change in the shape of the land which is enormous. For nothing now can float northward from the United States into Baffin's Bay. The polar current sets *out* of Baffin's Bay southward, bringing icebergs down, not leaves up, through Davis's Straits. And in any case we must allow that the hills of Disco Island were then the bottom of a sea: or how would the leaves have been deposited in them at all?

So much for the change of climate and land which can be proved to have gone on in Greenland. It has become colder. Why should it not some day become warmer again?

Now for England. It can be proved, as far as common sense can prove anything, that England was, before the age of ice, much warmer than it is now, and grew gradually cooler and cooler, just as, while the age of ice was dying out, it grew warmer again.

Now what proof is there of that?

This. Underneath London—as, I dare say, many of you know—there lies four or five hundred feet of clay. But not ice-clay. Anything but that, as you will see. It belongs to a formation late (geologically speaking), but somewhat older than those Disco Island beds.

And what sort of fossils do we find in it?

In the first place, the shells, which are abundant, are tropical—Nautili, Cones, and such like. And more, fruits and seeds are found in it, especially at the Isle of Sheppey. And what are they? Fruits of Nipa palms, a form only found now at river-mouths in Eastern India and the Indian islands; Anona-seeds; gourd-seeds; Acacia fruits—all tropical again; and Protea-ceous plants too—of an Australian type. Surely your common sense would hint to you, that this London clay must be mud laid down off the mouth of a tropical river. But your common sense would be all but certain of that, when you found, as you would find, the teeth and bones of crocodiles and turtles, who come to land, remember, to lay their eggs; the bones, too, of large mammals, allied to the tapir of India and South America, and the water-hog of the Cape. If all this does not mean that there was once a tropic climate and a tropic river running into some sea or other where London now stands, I must give up common sense and reason as deceitful and useless faculties; and believe nothing, not even the evidence of my own senses.

And now, have I, or have I not, fulfilled the promise which I made—rashly, I dare say some of you thought—in my first paper? Have I, or have I not, made you prove to yourself, by your own common sense, that the lowlands of Britain were underneath the sea in the days in which these pebbles and boulders were laid down over your plains? Nay, have we not proved more? Have we not found that that old sea was an icy sea? Have we not wandered on, step by step, into a whole true fairyland of wonders? to a time when all England, Scotland, and Ireland were as Greenland is now? when mud streams have rushed down from under glaciers on to a cold sea-bottom, when "ice, mast high, came floating by, as green as emerald?" when Snowdon was sunk for at least fourteen hundred feet of its height? when (as I could prove to you, had I time) the peaks of the highest Cumberland and Scotch mountains alone stood out, as islets in a frozen sea?

We want to get an answer to one strange question, and we have found a group of questions stranger still, and got them answered too. But so it is always in science. We know not what we shall discover. But this, at least, we know, that it will be far more wonderful than we had dreamed. The scientific explorer is always like Saul of old, who set out simply to find his father's asses, and found them—and a kingdom besides.

I should have liked to have told you more about this bygone age of ice. I should have liked to say something to you on the curious question—which is still an open one—whether there were not two ages of ice; whether the climate here did not, after perhaps thousands of years of Arctic cold, soften somewhat for a while—a few thousand years, perhaps—and then harden again into a second age of ice, somewhat less severe, probably, than the first. I should have liked to have hinted at the probable causes of this change—indeed, of the age of ice altogether—whether it was caused by a change in the distribution of land and water, or by change in the height and size of these islands, which made them large enough, and high enough, to carry a sheet of eternal snow inland; or whether, finally, the age of ice was caused by an actual change in the position of the whole planet with regard to its orbit round the sun—shifting at once the poles and the tropics; a deep question that latter, on which astronomers, whose business it is, are still at work, and on which, ere young folk are old, they will have discovered, I expect, some startling facts. On that last question, I, being no astronomer, cannot speak. But I should have liked to have said somewhat on matters on which I have knowledge enough, at least, to teach you how much there is to be learnt. I should have liked to tell the student of sea-animals—how the ice-age helps to explain, and is again explained by, the remarkable discoveries which Dr. Carpenter and Mr. Wyville Thompson have just made, in the deep-sea dredgings in the North Atlantic. I should have liked to tell the botanist somewhat of the pro-glacial flora—the plants which lived here before the ice, and lasted, some of them at least, through all those ages of fearful cold, and linger still on the summits of Snowdon, and the highest peaks of Cumberland and Scotland. I should have liked to have told the lovers of zoology about the animals which lived before the ice—of the mammoth, or woolly elephant; the woolly rhinoceros, the cave lion and bear, the reindeer, the musk oxen, the lemmings and the marmots which inhabited Britain till the ice drove them out southward, even into the South of France;

and how as the ice retreated, and the climate became toler-
able once more, some of them—the mammoth and rhinoceros,
the bison, the lion, and many another mighty beast reoccupied
our lowlands, at a time when the hippopotamus, at least in
summer, ranged freely from Africa and Spain across what was
then dry land between France and England, and fed by the
side of animals which have long since retreated to Norway and
to Canada. I should have liked to tell the archæologist of the
human beings—probably from their weapons and their habits—of the same race as the present Laplanders, who passed
northward as the ice went back, following the wild reindeer
herds from the South of France into our islands, which were
no islands then, to be in their turn driven northward by stronger races from the east and south. But space presses, and I fear
that I have written too much already.

At least, I have turned over for you a few grand and strange
pages in the book of nature, and taught you, I hope, a key by
which to decipher their hieroglyphics. At least, I have, I trust,
taught you to look, as I do, with something of interest, even of
awe, upon the pebbles in the street.

## III. THE STONES IN THE WALL

This is a large subject. For in the different towns of these islands, the walls are built of stones of almost every age, from the earliest to the latest; and the town-geologist may find a quite different problem to solve in the nearest wall, on moving from one town to another twenty miles off. All I can do, therefore, is to take one set of towns, in the walls of which one sort of stones is commonly found, and talk of them; taking care, of course, to choose a stone which is widely distributed. And such, I think, we can find in the so-called New Red sandstone, which, with its attendant marls, covers a vast tract—and that a rich and busy one—of England. From Hartlepool and the mouth of the Tees, down through Yorkshire and Nottinghamshire; over the manufacturing districts of central England; down the valley of the Severn; past Bristol and the Somersetshire flats to Torquay in South Devon; up north-westward through Shropshire and Cheshire; past Liverpool and northward through Lancashire; reappearing again, north of the Lake mountains, about Carlisle and the Scotch side of the Solway Frith, stretches the New Red sandstone plain, from under which everywhere the coal-bearing rocks rise as from a sea. It contains, in many places, excellent quarries of building-stone; the most famous of which, perhaps, are the well-known Runcorn quarries, near Liverpool, from which the old Romans brought the material for the walls and temples of ancient Chester, and from which the stone for the restoration of Chester Cathedral is being taken at this day.

In some quarters, especially in the north-west of England, its soil is poor, because it is masked by that very boulder-clay of which I spoke in my last paper. But its rich red marls, wherever they come to the surface, are one of God's most precious gifts to this favoured land. On them, one finds oneself at once in a garden; amid the noblest of timber, wheat, roots, grass which is green through the driest summers, and, in the western counties, cider-orchards laden with red and golden fruit. I know, throughout northern Europe, no such charming scenery, for quiet beauty and solid wealth, as that of the New Red marls; and if I wished to show a foreigner what England was, I should take him along them, from Yorkshire to South Devon, and say—There. Is not that a country worth living for,—and worth dying for if need be?

Another reason which I have for dealing with the New Red sandstone is this—that (as I said just now) over great tracts of England, especially about the manufacturing districts, the town-geologist will find it covered immediately by the boulder clay.

The townsman, finding this, would have a fair right to suppose that the clay was laid down immediately, or at least soon after, the sandstones or marls on which it lies; that as soon as the one had settled at the bottom of some old sea, the other settled on the top of it, in the same sea.

A fair and reasonable guess, which would in many cases, indeed in most, be quite true. But in this case it would be a mistake. The sandstone and marls are immensely older than the boulder-clay. They are, humanly speaking, some four or five worlds older.

What do I mean? This—that between the time when the one, and the time when the other, was made, the British Islands, and probably the whole continent of Europe, have changed four or five times; in shape; in height above the sea, or depth below it; in climate; in the kinds of plants and animals which have dwelt on them, or on their sea-bottoms. And surely it is not too strong a metaphor, to call such changes a change from an old world to a new one.

Mind. I do not say that these changes were sudden or violent. It is far more probable that they are only part and parcel of that vast but slow change which is going on everywhere over our whole globe. I think that will appear probable in the course of this paper. But that these changes have taken place, is my main thesis. The fact I assert; and I am bound to try and prove it. And in trying to do so, I shall no longer treat my readers, as I did in the first two papers, like children. I shall take for granted that they now understand something of the method by which geological problems are worked out; and can trust it, and me; and shall state boldly the conclusions of geologists, only giving proof where proof is specially needed.

Now you must understand that in England there are two great divisions of these New Red sandstones, "Trias," as geologists call them. An upper, called in Germany Keuper, which consists, atop, of the rich red marl, below them, of sandstones, and of those vast deposits of rock-salt, which have been long worked, and worked to such good purpose, that a vast subsidence of land has just taken place near Nantwich in Cheshire; and serious fears are entertained lest the town itself may subside, to fill up the caverns below, from whence the salt has been quarried. Underneath these beds again are those which carry the building-stone of Runcorn. Now these beds altogether, in Cheshire, at least, are about 3,400 feet thick; and were not laid down in a year, or in a century either.

Below them lies a thousand feet of sandstones, known in Germany by the name of "Bunter," from its mottled and spotted appearance. What lies under them again, does not concern us just now.

I said that the geologists called these beds the Trias; that is, the triple group. But as yet we have heard of only two parts of it. Where is the third?

Not here, but in Germany. There, between the Keuper above and the Bunter below, lies a great series of limestone beds, which, from the abundance of fossils which they contain, go by the name of Muschelkalk. A long epoch must therefore have intervened between the laying down of the Bunter and of the Keuper. And we have a trace of that long epoch, even in England. The Keuper lies, certainly, immediately on the Bunter; but not always "conformably" on it. That is, the beds are not exactly parallel. The Bunter had been slightly tilted, and slightly waterworn, before the Keuper was laid on it.

It is reasonable, therefore, to suppose, that the Bunter in England was dry land, and therefore safe from fresh deposit, through ages during which it was deep enough beneath the sea in Germany, to have the Muschelkalk laid down on it. Here again, then, as everywhere, we have evidence of time—time, not only beyond all counting, but beyond all imagining.

And now, perhaps, the reader will ask—If I am to believe that all new land is made out of old land, and that all rocks and soils are derived from the wear and tear of still older rocks, off what land came this enormous heap of sands more than 5,000 feet thick in places, stretching across England and into Germany?

It is difficult to answer. The shape and distribution of land in those days were so different from what they are now, that the rocks which furnished a great deal of our sandstone may be now, for aught I know, a mile beneath the sea.

But over the land which still stands out of the sea near us there has been wear and tear enough to account for any quantity of sand deposit. As a single instance—It is a provable and proven fact—as you may see from Mr. Ramsay's survey of North Wales—that over a large tract to the south of Snowdon, between Port Madoc and Barmouth, there has been ground off and carried away a mass of solid rock 20,000 feet thick; thick enough, in fact, if it were there still, to make a range of mountains as high as the Andes. It is a provable and proven fact that vast tracts of the centre of poor old Ireland were once covered with coal-measures, which have been scraped off in likewise, deprived of inestimable mineral wealth. The destruction of rocks—"denudation" as it is called—in the district round Malvern, is, I am told, provably enormous. Indeed, it is so over all Wales, North England, and West and North Scotland. So there is enough of rubbish to be accounted for to make our New Red sands. The round pebbles in it being, I believe, pieces of Old Red sandstone, may have come from the great Old Red sandstone region of South East Wales and Herefordshire. Some of the rubbish, too, may have come from what is now the Isle of Anglesey.

For you find in the beds, from the top to the bottom (at least in Cheshire), particles of mica. Now this mica could not have been formed in the sand. It is a definite crystalline mineral, whose composition is well known. It is only found in rocks which have been subjected to immense pressure, and probably to heat. The granites and mica-slates of Anglesey are full of it; and from Anglesey—as likely as from anywhere else—these thin scales of mica came. And that is about all that I can say on the matter. But it is certain that most of these sands were deposited in a very shallow water, and very near to land. Sand and pebbles, as I said in my first paper, could not be carried far out to sea; and some of the beds of the Bunter are full of rounded pebbles. Nay, it is certain that their surface was often out of water. Of that you may see very pretty proofs. You find these sands ripple-marked, as you do shore-sands now. You find cracks where the marl mud has dried in the sun: and, more, you find the little pits made by rain. Of that I have no doubt. I have seen specimens, in which you could not only see at a glance that the marks had been made by the large drops of a shower, but see also from what direction the shower had come. These delicate markings must have been covered up immediately with a fresh layer of mud or sand. How long since? How long since that flag had seen the light of the sun, when it saw it once again, restored to the upper air by the pick of the quarryman? Who can answer that? Not I.

Fossils are very rare in these sands; it is not easy to say why. It may be that the red oxide of iron in them has destroyed them. Few or none are ever found in beds in which it abounds. It is curious, too, that the Keuper, which is all but barren of fossils in England, is full of them in Wurtemberg, reptiles, fish, and remains of plants being common. But what will interest the reader are the footprints of a strange beast, found alike in England and in Germany—the Cheirotherium, as it was first named, from its hand-like feet; the Labyrinthodon, as it is now named, from the extraordinary structure of its teeth. There is little doubt now, among anatomists, that the bones and teeth of the so-called Labyrinthodon belong to the animal which made the footprints. If so, the creature must have been a right loathly monster. Some think him to have been akin to lizards; but the usual opinion is that he was a cousin of frogs and toads. Looking at his hands and other remains, one pictures him to oneself as a short, squat brute, as big as a fat hog, with a head very much the shape of a baboon, very large hands behind and small ones in front, waddling about on the tide flats of a sandy sea, and dragging after him, seemingly, a short tail, which has left its mark on the sand. What his odour was, whether he was smooth or warty, what he ate, and in general how he got his living, we know not. But there must have been something there for him to eat; and I dare say that he was about as happy and about as intellectual as the toad is now. Remember always that there is nothing alive now exactly like him, or, indeed, like any animal found in these sandstones. The whole animal world of this planet has changed entirely more than once since the Labyrinthodon waddled over the Cheshire flats. A lizard, for instance, which has been found in the Keuper, had a skull like a bird's, and no teeth—a type which is now quite extinct. But there is a more remarkable animal of which I must say a few words, and one which to scientific men is most interesting and significant.

Both near Warwick, and near Elgin in Scotland, in Central India, and in South Africa, fossil remains are found of a family of lizards utterly unlike anything now living save one, and that one is crawling about, plentifully I believe—of all places in the world—in New Zealand. How it got there; how so strange a type of creature should have died out over the rest of the world, and yet have lasted on in that remote island for long ages, ever since the days of the New Red sandstone, is one of those questions—quite awful questions I consider them—with which I will not puzzle my readers. I only mention it to show them what serious questions the scientific man has to face, and to answer, if he can. Only the next time they go to the Zoological Gardens in London, let them go to the reptile-house, and ask the very clever and courteous attendant to show them the Sphenodons, or Hatterias, as he will probably call them—and then look, I hope with kindly interest, at the oldest Conservatives they ever saw, or are like to see; gentlemen of most ancient pedigree, who have remained all but unchanged, while the whole surface of the globe has changed around them more than once or twice.

And now, of course, my readers will expect to hear something of the deposits of rock-salt, for which Cheshire and its red rocks are famous. I have never seen them, and can only say that the salt does not, it is said by geologists, lie in the sandstone, but at the bottom of the red marl which caps the sandstone. It was formed most probably by the gradual drying up of lagoons, such as are depositing salt, it is said now, both in the Gulf of Tadjara, on the Abyssinian frontier opposite Aden, and in the Runn of Cutch, near the Delta of the Indus. If this be so, then these New Red sandstones may be the remains of a whole Sahara—a sheet of sandy and all but lifeless deserts, reaching from the west of England into Germany, and rising slowly out of the sea; to sink, as we shall find, beneath the sea again.

And now, as to the vast period of time—the four or five worlds, as I called it—which elapsed between the laying down of the New Red sandstones and the laying down of the boulder-clays.

I think this fact—for fact it is—may be better proved by taking readers an imaginary railway journey to London from any spot in the manufacturing districts of central England—begging them, meanwhile, to keep their eyes open on the way.

And here I must say that I wish folks in general would keep their eyes a little more open when they travel by rail. When I see young people rolling along in a luxurious carriage, their eyes and their brains absorbed probably in a trashy shilling novel, and never lifted up to look out of the window, unconscious of all that they are passing—of the reverend antiquities, the admirable agriculture, the rich and peaceful scenery, the like of which no country upon earth can show; unconscious, too, of how much they might learn of botany and zoology, by simply watching the flowers along the railway banks and the sections in the cuttings: then it grieves me to see what little use people make of the eyes and of the understanding which God has given them. They complain of a dull journey: but it is not the journey which is dull; it is they who are dull. Eyes have they, and see not; ears have they, and hear not; mere dolls in smart clothes, too many of them, like the idols of the heathen.

But my readers, I trust, are of a better mind. So the next time they find themselves running up southward to London—or the reverse way—let them keep their eyes open, and verify, with the help of a geological map, the sketch which is given in the following pages.

Of the "Black Countries"—the actual coal districts I shall speak hereafter. They are in England either shores or islands yet undestroyed, which stand out of the great sea of New Red sandstone, and often carry along their edges layers of far younger rocks, called now Permian, from the ancient kingdom of Permia, in Russia, where they cover a vast area. With them I will not confuse the reader just now, but will only ask him to keep his eye on the rolling plain of New Red sands and marls past, say, Birmingham and Warwick. After those places, these sands and marls dip to the south-east, and other rocks and soils appear above them, one after another, dipping likewise towards the south-east—that is, toward London.

First appear thin layers of a very hard blue limestone, full of shells, and parted by layers of blue mud. That rock runs in a broad belt across England, from Whitby in Yorkshire, to Lyme in Dorsetshire, and is known as Lias. Famous it is, as some readers may know, for holding the bones of extinct monsters— Ichthyosaurs and Plesiosaurs, such as the unlearned may behold in the lake at the Crystal Palace. On this rock lie the rich cheese pastures, and the best tracts of the famous "hunting shires" of England.

Lying on it, as we go south-eastward, appear alternate beds of sandy limestone, with vast depths of clay between them. These "oolites," or freestones, furnish the famous Bath stone, the Oxford stone, and the Barnack stone of Northamptonshire, of which some of the finest cathedrals are built—a stone only surpassed, I believe, by the Caen stone, which comes from beds of the same age in Normandy. These freestones and clays abound in fossils, but of kinds, be it remembered, which differ more and more from those of the lias beneath, as the beds are higher in the series, and therefore nearer. There, too, are found principally the bones of that extraordinary flying lizard, the Pterodactyle, which had wings formed out of its fore-legs, on somewhat the same plan as those of a bat, but with one exception. In the bat, as any one may see, four fingers of the hand are lengthened to carry the wing, while the first alone is left free, as a thumb: but in the Pterodactyle, the outer or "little" finger alone is lengthened, and the other four fingers left free—one of those strange instances in nature of the same effect being produced in widely different plants and animals, and yet by slightly different means, on which a whole chapter of natural philosophy—say, rather, natural theology—will have to be written some day.

But now consider what this Lias, and the Oolites and clays upon it mean. They mean that the New Red sandstone, after it had been dry land, or all but dry land (as is proved by the footprints of animals and the deposits of salt), was sunk again beneath the sea. Each deposit of limestone signifies a long period of time, during which that sea was pure enough to allow reefs of coral to grow, and shells to propagate, at the bottom. Each great band of clay signifies a long period, during which fine mud was brought down from some wasting land in the neighbourhood. And that land was not far distant is proved by the bones of the Pterodactyle, of Crocodiles, and of Marsupials; by the fact that the shells are of shallow-water or shore species; by the presence, mixed with them, of fragments of wood, impressions of plants, and even wing-shells of beetles; and lastly, if further proof was needed, by the fact that in the "dirt-bed" of the Isle of Portland and the neighbouring shores, stumps of trees allied to the modern sago-palms are found as they grew in the soil, which, with them, has been covered up in layers of freshwater shale and limestone. A tropic forest has plainly sunk beneath a lagoon; and that lagoon, again, beneath the sea.

And how long did this period of slow sinking go on? Who can tell? The thickness of the Lias and Oolites together cannot be less than a thousand feet. Considering, then, the length of time required to lay down a thousand feet of strata, and considering the vast difference between the animals found in them, and the few found in the New Red sandstone, we have a right to call them another world, and that one which must have lasted for ages.

After we pass Oxford, or the Vale of Aylesbury, we enter yet another world. We come to a bed of sand, under which the freestones and their adjoining clays dip to the south-east. This is called commonly the lower Greensand, though it is not green, but rich iron-red. Then succeeds a band of stiff blue clay, called the Gault, and then another bed of sand, the upper Greensand, which is more worthy of the name, for it does carry, in most places, a band of green or "glauconite" sand. But it and the upper layers of the lower Greensand also, are worth our attention; for we are all probably eating them from time to time in the form of bran.

It had been long remarked that certain parts of these beds carried admirable wheatland; it had been remarked, too, that the finest hop-lands—those of Farnham, for instance, and Tunbridge—lay upon them: but that the fertile band was very narrow; that, as in the Surrey Moors, vast sheets of the lower Greensand were not worth cultivation. What caused the striking difference?

My beloved friend and teacher, the late Dr. Henslow, when Professor of Botany at Cambridge, had brought to him by a farmer (so the story ran) a few fossils. He saw, being somewhat of a geologist and chemist, that they were not, as fossils usually are, carbonate of lime, but phosphate of lime—bone-earth. He

said at once, as by an inspiration, "You have found a treasure—
not a gold-mine, indeed, but a food-mine. This is bone-earth,
which we are at our wits' end to get for our grain and pulse;
which we are importing, as expensive bones, all the way from
Buenos Ayres. Only find enough of them, and you will increase
immensely the food supply of England, and perhaps make her
independent of foreign phosphates in case of war."

His advice was acted on; for the British farmer is by no means
the stupid personage which townsfolk are too apt to fancy him.
This bed of phosphates was found everywhere in the Greensand, underlying the Chalk. It may be traced from Dorsetshire through England to Cambridge, and thence, I believe,
into Yorkshire. It may be traced again, I believe, all round the
Weald of Kent and Sussex, from Hythe to Farnham—where it
is peculiarly rich—and so to Eastbourne and Beachey Head;
and it furnishes, in Cambridgeshire, the greater part of those
so-called "coprolites," which are used perpetually now for manure, being ground up, and then treated with sulphuric acid,
till they become a "soluble super-phosphate of lime."

So much for the useless "hobby," as some fancy it, of poking
over old bones and stones, and learning a little of the composition of this earth on which God has placed us.

How to explain the presence of this vast mass of animal matter, in one or two thin bands right across England, I know not.
That the fossils have been rolled on a sea-beach is plain to
those who look at them. But what caused so vast a destruction
of animal life along that beach, must remain one of the buried
secrets of the past.

And now we are fast nearing another world, which is far younger than that coprolite bed, and has been formed under circum-
stances the most opposite to it. We are nearing, by whatever
rail we approach London, the escarpment of the chalk downs.

All readers, surely, know the white chalk, the special feature
and the special pride of the south of England. All know its
softly-rounded downs, its vast beech woods, its short and
sweet turf, its snowy cliffs, which have given—so some say—

to the whole island the name of Albion—the white land. But all do not, perhaps, know that till we get to the chalk no single plant or animal has been found which is exactly like any plant or animal now known to be living. The plants and animals grow, on the whole, more and more like our living forms as we rise in the series of beds. But only above the chalk (as far as we yet know) do we begin to find species identical with those living now.

This in itself would prove a vast lapse of time. We shall have a further proof of that vast lapse when we examine the chalk itself. It is composed—of this there is now no doubt—almost entirely of the shells of minute animalcules; and animalcules (I use an unscientific word for the sake of unscientific readers) like these, and in some cases identical with them, are now forming a similar deposit of mud, at vast depths, over the greater part of the Atlantic sea-floor. This fact has been put out of doubt by recent deep-sea dredgings. A whole literature has been written on it of late. Any reader who wishes to know it, need only ask the first geologist he meets; and if he has the wholesome instinct of wonder in him, fill his imagination with true wonders, more grand and strange than he is like to find in any fairy tale. All I have to do with the matter here is, to say that, arguing from the known to the unknown, from the Atlantic deep-sea ooze which we do know about, to the chalk which we do not know about, the whole of the chalk must have been laid down at the bottom of a deep and still ocean, far out of the reach of winds, tides, and even currents, as a great part of the Atlantic sea-floor is at this day.

Prodigious! says the reader. And so it is. Prodigious to think that that shallow Greensand shore, strewed with dead animals, should sink to the bottom of an ocean, perhaps a mile, perhaps some four miles deep. Prodigious the time during which it must have lain as a still ocean-floor. For so minute are the living atomies which form the ooze, that an inch, I should say, is as much as we can allow for their yearly deposit; and the chalk is at least a thousand feet thick. It may have taken, therefore, twelve thousand years to form the chalk alone. A rough guess, of course, but one as likely to be two or three times too little as two or three times too big. Such, or somewhat such, is the fact. It had long been suspected, and more than suspected; and the late discoveries of Dr. Carpenter and Mr. Wyville Thompson have surely placed it beyond doubt.

Thus, surely, if we call the Oolitic beds one new world above the New Red sandstone, we must call the chalk a second new world in like wise.

I will not trouble the reader here with the reasons why geologists connect the chalk with the greensands below it, by regular gradations, in spite of the enormous downward leap, from sea-shore to deep ocean, which the beds seem (but only seem) to have taken. The change—like all changes in geology— was probably gradual. Not by spasmodic leaps and starts, but slowly and stately, as befits a God of order, of patience, and of strength, have these great deeds been done.

But we have not yet done with new worlds or new prodigies on our way to London, as any Londoner may ascertain for himself, if he will run out a few miles by rail, and look in any cutting or pit, where the surface of the chalk, and the beds which lie on it, are exposed.

On the chalk lie—especially in the Blackheath and Woolwich district—sands and clays. And what do they tell us?

Of another new world, in which the chalk has been lifted up again, to form gradually, doubtless, and at different points in succession, the shore of a sea.

But what proof is there of this?

The surface of the chalk is not flat and smooth, as it must have been when at the bottom of the sea. It is eaten out into holes and furrows, plainly by the gnawing of the waves; and on it lie, in many places, large rolled flints out of chalk which has been destroyed, beds of shore-shingle, beds of oysters lying as they grew, fresh or brackish water-shells standing as they lived, bits of lignite (fossil wood half turned to coal), and (as in Katesgrove pits at Reading) leaves of trees. Proof enough, one would say, that the chalk had been raised till part of it at least became dry land, and carried vegetation.

And yet we have not done. There is another world to tell of yet.

For these beds (known as the Woolwich and Reading beds) dip under that vast bed of London clay, four hundred and more feet thick, which (as I said in my last chapter) was certainly laid down by the estuary of some great tropic river, among palm-trees and Anonas, crocodiles and turtles.

Is the reader's power of belief exhausted?

If not: there are to be seen, capping almost every high land round London, the remains of a fifth world. Some of my readers may have been to Ascot races, or to Aldershot camp, and may recollect the table-land of the sandy moors, perfectly flat atop, dreary enough to those to whom they are not (as they have long been to me) a home and a work-field. Those sands are several hundred feet thick. They lie on the London clay. And they represent—the reader must take geologists' word for it—a series of beds in some places thousands of feet thick, in the Isle of Wight, in the Paris basin, in the volcanic country of the Auvergne, in Switzerland, in Italy; a period during which the land must at first have swarmed with forms of tropic life, and then grown—but very gradually—more temperate, and then colder and colder still; till at last set in that age of ice, which spread the boulder pebbles over all rocks and soils indiscriminately, from the Lake mountains to within a few miles of London.

For everywhere about those Ascot moors, the top of the sands has been ploughed by shore-ice in winter, as they lay a-wash in the shallow sea; and over them, in many places, is spread a thin sheet of ice gravel, more ancient, the best geologists think, than the boulder and the boulder-clay.

If any of my readers ask how long the period was during which those sands of Ascot Heath and Aldershot have been laid down, I cannot tell. But this we can tell. It was long enough to see such changes in land and sea, that maps representing

Europe during the greater part of that period (as far as we can guess at it) look no more like Europe than like America or the South Sea Islands. And this we can tell besides: that that period was long enough for the Swiss Alps to be lifted up at least 10,000 feet of their present height. And that was a work which—though God could, if He willed it, have done it in a single day—we have proof positive was not done in less than ages, beside which the mortal life of man is as the life of the gnat which dances in the sun.

And all this, and more—as may be proved from the geology of foreign countries—happened between the date of the boulder-clay, and that of the New Red sandstone on which it rests.

# IV. THE COAL IN THE FIRE

My dear town-dwelling readers, let me tell you now something of a geological product well known, happily, to all dwellers in towns, and of late years, thanks to railroad extension, to most dwellers in country districts: I mean coal.

Coal, as of course you know, is commonly said to be composed of vegetable matter, of the leaves and stems of ancient plants and trees—a startling statement, and one which I do not wish you to take entirely on trust. I shall therefore spend a few pages in showing you how this fact—for fact it is—was discovered. It is a very good example of reasoning from the known to the unknown. You will have a right to say at first starting, "Coal is utterly different in look from leaves and stems. The only property which they seem to have in common is that they can both burn." True. But difference of mere look may be only owing to a transformation, or series of transformations. There are plenty in nature quite as great, and greater. What can be more different in look, for instance, than a green field of wheat and a basket of loaves at the baker's? And yet there is, I trust, no doubt whatsoever that the bread has been once green wheat, and that the green wheat has been transformed into bread—making due allowance, of course, for the bone-dust, or gypsum, or alum with which the worthy baker may have found it profitable to adulterate his bread, in order to improve the digestion of Her Majesty's subjects.

But you may say, "Yes, but we can see the wheat growing, flow-ering, ripening, reaped, ground, kneaded, baked. We see, in the case of bread, the processes of the transformation going on: but in the case of coal we do not see the wood and leaves being actually transformed into coal, or anything like it."

Now suppose we laid out the wheat on a table in a regular se-ries, such as you may see in many exhibitions of manufactures; beginning with the wheat plant at one end, and ending with the loaf at the other; and called in to look at them a savage who knew nothing of agriculture and nothing of cookery—called in, as an extreme case, the man in the moon, who certainly can know nothing of either; for as there is neither air nor wa-ter round the moon, there can be nothing to grow there, and therefore nothing to cook—and suppose we asked him to study the series from end to end. Do you not think that the man in the moon, if he were half as shrewd as Crofton Croker makes him in his conversation with Daniel O'Rourke, would answer after due meditation, "How the wheat plant got changed into the loaf I cannot see from my experience in the moon: but that it has been changed, and that the two are the same thing I do see, for I see all the different stages of the change." And so I think you may say of the wood and the coal.

The man in the moon would be quite reasonable in his conclu-sion; for it is a law, a rule, and one which you will have to apply again and again in the study of natural objects, that however different two objects may look in some respects, yet if you can find a regular series of gradations between them, with all shades of likeness, first to one of them and then to the other, then you have a fair right to suppose them to be only varieties of the same species, the same kind of thing, and that, there-fore, they have a common origin.

That sounds rather magniloquent. Let me give you a simple example.

Suppose you had come into Britain with Brute, the grandson of Æneas, at that remote epoch when (as all archaeologists know who have duly read Geoffrey of Monmouth and the Arthuric legends) Britain was inhabited only by a few giants. Now if you had met giants with one head, and also giants with seven heads, and no others, you would have had a right to say, "There are two breeds of giants here, one-headed and seven-headed." But if you had found, as Jack the Giant-Killer (who belongs to the same old cycle of myths) appears to have found, two-headed giants also, and three-headed, and giants, indeed, with any reasonable number of heads, would you not have been justified in saying, "They are all of the same breed, after all; only some are more capitate, or heady, than others!"

I hope that you agree to that reasoning; for by it I think we arrive most surely at a belief in the unity of the human race, and that the Negro is actually a man and a brother.

If the only two types of men in the world were an extreme white type, like the Norwegians, and an extreme black type, like the Negros, then there would be fair ground for saying, "These two types have been always distinct; they are different races, who have no common origin." But if you found, as you will find, many types of man showing endless gradations between the white man and the Negro, and not only that, but endless gradations between them both and a third type, whose extreme perhaps is the Chinese—endless gradations, I say, showing every conceivable shade of resemblance or difference, till you often cannot say to what type a given individual belongs; and all of them, however different from each other, more like each other than they are like any other creature upon earth; then you are justified in saying, "All these are mere varieties of one kind. However distinct they are now, they were probably like each other at first, and therefore all probably had a common origin." That seems to me sound reasoning, and advanced natural science is corroborating it more and more daily.

Now apply the same reasoning to coal. You may find about the world—you may see even in England alone—every gradation between coal and growing forest. You may see the forest growing in its bed of vegetable mould; you may see the forest dead and converted into peat, with stems and roots in it; that, again, into sunken forests, like those to be seen below high-water mark on many coasts of this island. You find gradations between them and beds of lignite, or wood coal; then gradations between lignite and common or bituminous coal; and then gradations between common coal and culm, or anthracite, such as is found in South Wales. Have you not a right to say, "These are all but varieties of the same kind of thing—namely, vegetable matter? They have a common origin—namely, woody fibre. And coal, or rather culm, is the last link in a series of transformations from growing vegetation?"

This is our first theory. Let us try to verify it, as scientific men are in the habit of doing, by saying, If that be true, then something else is likely to be true too.

If coal has all been vegetable soil, then it is likely that some of it has not been quite converted into shapeless coal. It is likely that there will be vegetable fibre still to be seen here and there; perhaps leaves, perhaps even stems of trees, as in a peat bog. Let us look for them.

You will not need to look far. The coal, and the sands and shales which accompany the coal, are so full of plant-remains, that three hundred species were known to Adolphe Brongniart as early as 1849, and that number has largely increased since.

Now one point is specially noticeable about these plants of the coal; namely, that they may at least have grown in swamps.

First, you will be interested if you study the coal flora, with the abundance, beauty, and variety of the ferns. Now ferns in these islands grow principally in rocky woods, because there, beside the moisture, they get from decaying vegetable or decaying rock, especially limestone, the carbonic acid which is their special food, and which they do not get on our dry pastures, and still less in our cultivated fields. But in these islands there are two noble species, at least, which are true swamp-ferns; the Lastræa Thelypteris, which of old filled the fens, but is now all but extinct; and the Osmunda, or King-fern, which, as all know, will grow wherever it is damp enough about the roots. In Hampshire, in Devon, and Cornwall, and in the southwest of Ireland, the King-fern too is a true swamp fern. But in the

Tropics I have seen more than once noble tree-ferns growing in wet savannahs at the sea-level, as freely as in the mountain-woods; ferns with such a stem as some of the coal ferns had, some fifteen feet in height, under which, as one rode on horseback, one saw the blazing blue sky, as through a parasol of delicate lace, as men might have long ages since have seen it, through the plumed fronds of the ferns now buried in the coal, had there only been a man then created to enjoy its beauty.

Next we find plants called by geologists Calamites. There is no doubt now that they are of the same family as our Equiseta, or horse-tails, a race which has, over most parts of the globe, dwindled down now from twenty or thirty feet in height, as they were in the old coal measures, to paltry little weeds. The tallest Equisetum in England—the beautiful E. Telmateia—is seldom five feet high. But they, too, are mostly mud and swamp plants; and so may the Calamites have been.

The Lepidodendrons, again, are without doubt the splendid old representatives of a family now dwindled down to such creeping things as our club-mosses, or Lycopodiums. Now it is a certain fact, which can be proved by the microscope, that a very great part of the best coal is actually made up of millions of the minute seeds of club-mosses, such as grow—a few of them, and those very small—on our moors; a proof, surely, not only of the vast amount of the vegetation in the coal-making age, but also of the vast time during which it lasted. The Lepidodendra may have been fifty or sixty feet high. There is not a Lycopodium in the world now, I believe, five feet high. But the club-mosses are now, in these islands and elsewhere, lovers of wet and peaty soils, and so may their huger prototypes have been, in the old forests of the coal.

Of the Sigillariæ we cannot say as much with certainty, for botanists are not agreed as to what low order of flowerless plants they belong. But that they rooted in clay beds there is proof, as you will hear presently.

And as to the Conifers, or pine-like trees—the Dadoxylon, of which the pith goes by the name of Sternbergia, and the uncertain tree which furnishes in some coal-measures bushels of a seed connected with that of the yew—we may suppose that they would find no more difficulty in growing in swamps than the cypress, which forms so large a portion of the vegetation in the swamps of the Southern United States.

I have given you these hints, because you will naturally wish to know what sort of a world it was in which all these strange plants grew and turned into coal.

My answer is, that it was most probably just like the world in which we are living now, with the one exception that the plants and animals are different.

It was the fashion a few years since to explain the coal—like other phenomena of geology—by some mere hypothesis of a state of things quite unlike what we see now. We were brought up to believe that in the Carboniferous, or coal-bearing era, the atmosphere was intensely moist and hot, and overcharged with carbonic acid, which had been poured out from the interior of the planet by volcanic eruptions, or by some other convulsion. I forget most of it now: and really there is no need to remember; for it is all, I verily believe, a dream—an attempt to explain the unknown not by the known, but by the still more unknown. You may find such theories lingering still in sensational school-books, if you like to be unscientific. If you like, on the other hand, to be scientific you will listen to those who tell you that instead of there having been one unique carboniferous epoch, with a peculiar coal-making climate, all epochs are carboniferous if they get the chance; that coal is of every age, from that of the Scotch and English beds, up to the present day. The great coal-beds along the Rocky Mountains, for instance, are tertiary—that is, later than the chalk. Coal is forming now, I doubt not, in many places on the earth, and would form in many more, if man did not interfere with the processes of wild nature, by draining the fens, and embanking the rivers.

Let me by a few words prove this statement. They will give you, beside, a fresh proof of Sir Charles Lyell's great geological rule—that the best way to explain what we see in ancient rocks is to take for granted, as long as we can do so fairly, that things were going on then very much as they are going on now.

When it was first seen that coal had been once vegetable, the question arose—How did all these huge masses of vegetable matter get there? The Yorkshire and Derbyshire coal-fields, I hear, cover 700 or 800 square miles; the Lancashire about 200. How large the North Wales and the Scotch fields are I cannot say. But doubtless a great deal more coal than can be got at lies under the sea, especially in the north of Wales. Coal probably exists over vast sheets of England and France, buried so deeply under later rocks, that it cannot be reached by mining. As an instance, a distinguished geologist has long held that there are beds of coal under London itself, which rise, owing to a peculiar disturbance of the strata, to within 1,000 or 1,200 feet of the surface, and that we or our children may yet see coal-mines in the marshes of the Thames. And more, it is a provable fact that only a portion of the coal measures is left. A great part of Ireland must once have been covered with coal, which is now destroyed. Indeed, it is likely that the coal now known of in Europe and America is but a remnant of what has existed there in former ages, and has been eaten away by the inroads of the sea.

Now whence did all that enormous mass of vegetable soil come? Off some neighbouring land, was the first and most natural answer. It was a rational one. It proceeded from the known to the unknown. It was clear that these plants had grown on land; for they were land-plants. It was clear that there must have been land close by, for between the beds of coal, as you all know, the rock is principally coarse sandstone, which could only have been laid down (as I have explained to you already) in very shallow water.

It was natural, then, to suppose that these plants and trees had been swept down by rivers into the sea, as the sands and muds which buried them had been. And it was known that at the mouths of certain rivers—the Mississippi, for instance— vast rafts of dead floating trees accumulated; and that the bottoms of the rivers were often full of snags, etc.; trees which had grounded, and stuck in the mud; and why should not the coal have been formed in the same way?

Because—and this was a serious objection—then surely the coal would be impure—mixed up with mud and sand, till it was not worth burning. Instead of which, the coal is usually pure vegetable, parted sharply from the sandstone which lies on it. The only other explanation was, that the coal vegetation had grown in the very places where it was found. But that seemed too strange to be true, till that great geologist, Sir W. Logan—who has since done such good work in Canada—showed that every bed of coal had a bed of clay under it, and that that clay always contained fossils called Stigmaria. Then it came out that the Stigmaria in the under clay had long filaments attached to them, while when found in the sandstones or shales, they had lost their filaments, and seemed more or less rolled—in fact, that the natural place of the Stigmaria was in the under clay. Then Mr. Binney discovered a tree—a Sigillaria, standing upright in the coal-measures with its roots attached. Those roots penetrated into the under clay of the coal; and those roots were Stigmarias. That seems to have settled the question. The Sigillarias, at least, had grown where they were found, and the clay beneath the coal-beds was the original soil on which they had grown. Just so, if you will look at any peat bog you will find it bottomed by clay, which clay is pierced everywhere by the roots of the moss forming the peat, or of the trees, birches, alders, poplars, and willows, which grow in the bog. So the proof seemed complete, that the coal had been formed out of vegetation growing where it was buried. If any further proof for that theory was needed, it would be found in this fact, most ingeniously suggested by Mr. Boyd Dawkins. The resinous spores, or seeds of the Lepidodendra make up—as said above—a great part of the bituminous coal. Now those spores are so light, that if the coal had been laid down by water, they would have floated on it, and have been carried away; and therefore the bituminous coal must have been formed, not under water, but on dry land.

I have dwelt at length on these further arguments, because they seem to me as pretty a specimen as I can give my readers of that regular and gradual induction, that common-sense regulated, by which geological theories are worked out.

But how does this theory explain the perfect purity of the coal? I think Sir C. Lyell answers that question fully in p. 383 of his "Student's Elements of Geology." He tells us that the dense growths of reeds and herbage which encompass the margins of forest-covered swamps in the valley and delta of the Mississippi, in passing through them, are filtered and made to clear themselves entirely before they reach the areas in which vegetable matter may accumulate for centuries, forming coal if the climate be favourable; and that in the cypress-swamps of that region no sediment mingles with the vegetable matter accumulated from the decay of trees and semi-aquatic plants; so that when, in a very dry season, the swamp is set on fire, pits are burnt into the ground many feet deep, or as far as the fire can go down without reaching water, and scarcely any earthy residuum is left; just as when the soil of the English fens catches fire, red-hot holes are eaten down through pure peat till the water-bearing clay below is reached. But the purity of the water in peaty lagoons is observable elsewhere than in the delta of the Mississippi. What can be more transparent than many a pool surrounded by quaking bogs, fringed, as they are in Ireland, with a ring of white water-lilies, which you dare not stoop to pick, lest the peat, bending inward, slide you down into that clear dark gulf some twenty feet in depth, bottomed and walled with yielding ooze, from which there is no escape? Most transparent, likewise, is the water of the West Indian swamps. Though it is of the colour of coffee, or rather of dark beer, and so impregnated with gases that it produces fever or cholera when drunk, yet it is—at least when it does not mingle with the salt water—so clear, that one might see every marking on a boa-constrictor or alligator, if he glided along the bottom under the canoe.

But now comes the question—Even if all this be true, how were the forests covered up in shale and sandstone, one after another?

By gradual sinking of the land, one would suppose.

If we find, as we may find in a hundred coal-pits, trees rooted as they grew, with their trunks either standing up through the coal, and through the sandstone above the coal; their bark often remaining as coal while their inside is filled up with sandstone, has not our common-sense a right to say—The land on which they grew sank below the water-line; the trees were killed; and the mud and sand which were brought down the streams enveloped their trunks? As for the inside being full of sandstone, have we not all seen hollow trees? Do we not all know that when a tree dies its wood decays first, its bark last? It is so, especially in the Tropics. There one may see huge dead trees with their bark seemingly sound, and their inside a mere cavern with touchwood at the bottom; into which caverns one used to peep with some caution. For though one might have found inside only a pair of toucans, or parrots, or a whole party of jolly little monkeys, one was quite as likely to find a poisonous snake four or five feet long, whose bite would have very certainly prevented me having the pleasure of writing this book.

Now is it not plain that if such trees as that sunk, their bark would be turned into lignite, and at last into coal, while their insides would be silted up with mud and sand? Thus a core or pillar of hard sandstone would be formed, which might do to the collier of the future what they are too apt to do now in the Newcastle and Bristol collieries. For there, when the coal is worked out below, the sandstone stems—"coal-pipes" as the colliers call them—in the roof of the seam, having no branches, and nothing to hold them up but their friable bark of coal, are but too apt to drop out suddenly, killing or wounding the hapless men below.

Or again, if we find—as we very often find—as was found at Parkfield Colliery, near Wolverhampton, in the year 1814—a quarter of an acre of coal-seam filled. with stumps of trees as they grew, their trunks broken off and lying in every direction, turned into coal, and flattened, as coal-fossils so often are, by the weight of the rock above—should we not have a right to say—These trees were snapped off where they grew by some violent convulsion; by a storm, or by a sudden inrush of water owing to a sudden sinking of the land, or by the very earth-quake shock itself which sank the land?

But what evidence have we of such sinkings? The plain fact that you have coal-seam above coal-seam, each with its bed of under-clay; and that therefore the land MUST have sunk ere the next bed of soil could have been deposited, and the next forest have grown on it.

In one of the Rocky Mountain coal-fields there are more than thirty seams of coal, each with its under-clay below it. What can that mean but thirty or more subsidences of the land, and the peat of thirty or more forests or peat-mosses, one above the other? And now if any reader shall say, Subsidence? What is this quite new element which you have brought into your ar-gument? You told us that you would reason from the known to the unknown. What do we know of subsidence? You offered to explain the thing which had gone on once by that which is go-ing on now. Where is subsidence going on now upon the surface of our planet? And where, too, upheaval, such as would bring us these buried forests up again from under the sea-level, and make them, like our British coal-field, dry land once more?

The answer is—Subsidence and elevation of the land are com-mon now, probably just as common as they were in any age of this planet's history.

To give two instances, made now notorious by the writings of geologists. As lately as 1819 a single earthquake shock in Cutch, at the mouth of the Indus, sunk a tract of land larger than the Lake of Geneva in some places to a depth of eighteen feet, and converted it into an inland sea. The same shock raised, a few miles off, a corresponding sheet of land some fifty miles in length, and in some parts sixteen miles broad, ten feet above the level of the alluvial plain, and left it to be named by the country-people the "Ullah Bund," or bank of God, to distin-guish it from the artificial banks in the neighbourhood.

Again: in the valley of the Mississippi—a tract which is now, it would seem, in much the same state as central England was while our coal-fields were being laid down—the earthquakes of 1811-12 caused large lakes to appear suddenly in many parts of the district, amid the dense forests of cypress. One of these, the "Sunk Country," near New Madrid, is between seventy and eighty miles in length, and thirty miles in breadth, and throughout it, as late as 1846, "dead trees were conspicuous, some erect in the water, others fallen, and strewed in dense masses over the bottom, in the shallows, and near the shore." I quote these words from Sir Charles Lyell's "Principles of Geology" (11th edit.), vol. i. p. 453. And I cannot do better than advise my readers, if they wish to know more of the way in which coal was formed, to read what is said in that book concerning the Delta of the Mississippi, and its strata of forests sunk where they grew, and in some places upraised again, alternating with beds of clay and sand, vegetable soil, recent sea-shells, and what not, forming, to a depth of several hundred feet, just such a mass of beds as exists in our own coal-fields at this day.

If, therefore, the reader wishes to picture to himself the scenery of what is now central England, during the period when our coal was being laid down, he has only, I believe, to transport himself in fancy to any great alluvial delta, in a moist and warm climate, favourable to the growth of vegetation. He has only to conceive wooded marshes, at the mouth of great rivers, slowly sinking beneath the sea; the forests in them killed by the water, and then covered up by layers of sand, brought down from inland, till that new layer became dry land, to carry a fresh crop of vegetation. He has thus all that he needs to explain how coal-measures were formed. I myself saw once a scene of that kind, which I should be sorry to forget; for there was, as I conceived, coal, making, or getting ready to be made, before my eyes: a sheet of swamp, sinking slowly into the sea; for there stood trees, still rooted below high-water mark, and killed by the waves; while inland huge trees stood dying, or dead, from the water at their roots. But what a scene—a labyrinth of narrow creeks, so narrow that a canoe could not pass up, haunted with alligators and boa-constrictors, parrots and white herons, amid an inextricable confusion of vegetable mud, roots of the alder-like mangroves, and tangled creepers hanging from tree to tree; and overhead huge fan-palms, delighting in the moisture, mingled with still huger broad-leaved trees in every stage of decay. The drowned vegetable soil of ages beneath me; above my head, for a hundred feet, a mass of stems and boughs, and leaves and flowers, compared with which the richest hothouse in England was poor and small. But if the sinking process which was going on continued a few hundred years, all that huge mass of wood and leaf would be sunk beneath the swamp, and covered up in mud washed down from the mountains, and sand driven in from the sea; to form a bed many feet thick, of what would be first peat, then lignite, and last, it may be, coal, with the stems of killed trees standing up out of it into the new mud and sand-beds above it, just as the Sigillariæ and other stems stand up in the coal-beds both of Britain and of Nova Scotia; while over it a fresh forest would grow up, to suffer the same fate—if the sinking process went on—as that which had preceded it.

That was a sight not easily to be forgotten. But we need not have gone so far from home, at least, a few hundred years ago, to see an exactly similar one. The fens of Norfolk and Cambridgeshire, before the rivers were embanked, the water pumped off, the forests felled, and the reed-beds ploughed up, were exactly in the same state. The vast deposits of peat between Cambridge and the sea, often filled with timber-trees, either fallen or upright as they grew, and often mixed with beds of sand or mud, brought down in floods, were formed in exactly the same way; and if they had remained undrained, then that slow sinking, which geologists say is going on over the whole area of the Fens, would have brought them gradually, but surely, below the sea-level, to be covered up by new forests, and converted in due time into coal. And future geologists would have found—they may find yet, if, which God forbid, England should become barbarous and the trees be thrown out of cultivation—instead of fossil Lepidodendra and Sigillariæ, Calamites and ferns, fossil ashes and oaks, alders and poplars, bulrushes and reeds. Almost the only fossil fern would have been that tall and beautiful Lastræa Thelypteris, once so abundant, now all but destroyed by drainage and the plough.

We need not, therefore, fancy any extraordinary state of things on this planet while our English coal was being formed. The climate of the northern hemisphere—Britain, at least, and Nova Scotia—was warmer than now, to judge from the abundance of ferns; and especially of tree-ferns; but not so warm, to judge from the presence of conifers (trees of the pine tribe), as the Tropics. Moreover, there must have been, it seems to me, a great scarcity of animal-life. Insects are found, beautifully preserved; a few reptiles, too, and land-shells; but very few. And where are the traces of such a swarming life as would be entombed were a tropic forest now sunk; which is found entombed in many parts of our English fens? The only explanation which I can offer is this—that the club-mosses, tree-ferns, pines, and other low-ranked vegetation of the coal afforded little or no food for animals, as the same families of plants do to this day; and if creatures can get nothing to eat, they certainly cannot multiply and replenish the earth. But, be that as it may, the fact that coal is buried forest is not affected.

Meanwhile, the shape and arrangements of sea and land must have been utterly different from what they are now. Where was that great land, off which great rivers ran to deposit our coal-measures in their deltas? It has been supposed, for good reasons, that north-western France, Belgium, Holland, and Germany were then under the sea; that Denmark and Norway were joined to Scotland by a continent, a tongue of which ran across the centre of England, and into Ireland, dividing the northern and southern coal-fields. But how far to the west and north did that old continent stretch? Did it, as it almost certainly did long ages afterwards, join Greenland and North America with Scotland and Norway? Were the northern fields of Nova Scotia, which are of the same geological age as our own, and contain the same plants, laid down by rivers which ran off the same continent as ours? Who can tell now? That old land, and all record of it, save what these fragmentary coal-measures can give, are buried in the dark abyss of countless ages; and we can only look back with awe, and comfort ourselves with the thought—Let Time be ever so vast, yet Time is not Eternity.

One word more. If my readers have granted that all for which I have argued is probable, they will still have a right to ask for further proof.

They will be justified in saying: "You say that coal is transformed vegetable matter; but can you show us how the transformation takes place? Is it possible according to known natural laws?"

The chemist must answer that. And he tells us that wood can become lignite, or wood-coal, by parting with its oxygen, in the shape of carbonic acid gas, or choke-damp; and then common or bituminous coal, by parting with its hydrogen, chiefly in the form of carburetted hydrogen—the gas with which we light our streets. That is about as much as the unscientific reader need know. But it is a fresh corroboration of the theory that coal has been once vegetable fibre, for it shows how vegetable fibre can, by the laws of nature, become coal. And it certainly helps us to believe that a thing has been done, if we are shown that it can be done.

This fact explains, also, why in mines of wood-coal carbonic acid, *i.e.* choke-damp, alone is given off. For in the wood-coal a great deal of the hydrogen still remains. In mines of true coal, not only is choke-damp given off, but that more terrible pest of the miners, fire-damp, or explosive carburetted hydrogen and olefiant gases. Now the occurrence of that fire-damp in mines proves that changes are still going on in the coal: that it is getting rid of its hydrogen, and so progressing toward the state of anthracite or culm—stone-coal as it is sometimes called. In the Pennsylvanian coal-fields some of the coal has actually done this, under the disturbing force of earthquakes; for the coal, which is bituminous, like our common coal, to the westward where the strata are horizontal, becomes gradually anthracite as it is tossed and torn by the earthquake faults of the Alleghany and Appalachian mountains.

And is a further transformation possible? Yes; and more than one. If we conceive the anthracite cleared of all but its last atoms of oxygen, hydrogen, and nitrogen, till it has become all but pure carbon, it would become—as it has become in certain rocks of immense antiquity, graphite—what we miscall black-lead. And, after that, it might go through one transformation more, and that the most startling of all. It would need only perfect purification and crystallisation to become—a diamond; nothing less. We may consider the coal upon the fire as the middle term of a series, of which the first is live wood, and the last diamond; and indulge safely in the fancy that every diamond in the world has probably, at some remote epoch, formed part of a growing plant.

A strange transformation; which will look to us more strange, more truly poetical, the more steadily we consider it.

The coal on the fire; the table at which I write—what are they made of? Gas and sunbeams; with a small percentage of ash, or earthy salts, which need hardly be taken into account.

Gas and sunbeams. Strange, but true.

The life of the growing plant—and what that life is who can tell?—laid hold of the gases in the air and in the soil; of the carbonic acid, the atmospheric air, the water—for that too is gas. It drank them in through its rootlets: it breathed them in through its leaf-pores, that it might distil them into sap, and bud, and leaf, and wood. But it has to take in another element, without which the distillation and the shaping could never have taken place. It had to drink in the sunbeams—that mysterious and complex force which is for ever pouring from the sun, and making itself partly palpable to our senses as heat and light. So the life of the plant seized the sunbeams, and absorbed them, buried them in itself—no longer as light and heat, but as invisible chemical force, locked up for ages in that woody fibre.

So it is. Lord Lytton told us long ago, in a beautiful song, how

The Wind and the Beam loved the Rose.

But Nature's poetry was more beautiful than man's. The wind and the beam loved the rose so well that they made the rose—

or rather, the rose took the wind and the beam, and built up out of them, by her own inner life, her exquisite texture, hue, and fragrance.

What next? The rose dies; the timber tree dies; decays down into vegetable fibre, is buried, and turned to coal: but the plant cannot altogether undo its own work. Even in death and decay it cannot set free the sunbeams imprisoned in its tissue. The sun-force must stay, shut up age after age, invisible, but strong; working at its own prison-cells; transmuting them, or making them capable of being transmuted by man, into the manifold products of coal—coke, petroleum, mineral pitch, gases, coal-tar, benzole, delicate aniline dyes, and what not, till its day of deliverance comes.

Man digs it, throws it on the fire, a black, dead-seeming lump. A corner, an atom of it, warms till it reaches the igniting point; the temperature at which it is able to combine with oxygen.

And then, like a dormant live thing, awaking after ages to the sense of its own powers, its own needs, the whole lump is seized, atom after atom, with an infectious hunger for that oxygen which it lost centuries since in the bottom of the earth. It drinks the oxygen in at every pore; and burns.

And so the spell of ages is broken. The sun-force bursts its prison-cells, and blazes into the free atmosphere, as light and heat once more; returning in a moment into the same forms in which it entered the growing leaf a thousand centuries since.

Strange it all is, yet true. But of nature, as of the heart of man, the old saying stands—that truth is stranger than fiction.

## V. THE LIME IN THE MORTAR

I shall presume in all my readers some slight knowledge about lime. I shall take for granted, for instance, that all are better informed than a certain party of Australian black fellows were a few years since.

In prowling on the track of a party of English settlers, to see what they could pick up, they came—oh joy!—on a sack of flour, dropped and left behind in the bush at a certain creek. The poor savages had not had such a prospect of a good meal for many a day. With endless jabbering and dancing, the whole tribe gathered round the precious flour-bag with all the pannikins, gourds, and other hollow articles it could muster, each of course with a due quantity of water from the creek therein, and the chief began dealing out the flour by handfuls, beginning of course with the boldest warriors. But, horror of horrors, each man's porridge swelled before his eyes, grew hot, smoked, boiled over. They turned and fled, man, woman, and child, from before that supernatural prodigy; and the settlers coming back to look for the dropped sack, saw a sight which told the whole tale. For the poor creatures, in their terror, had thrown away their pans and calabashes, each filled with that which it was likely to contain, seeing that the sack itself had contained, not flour, but quick-lime. In memory of which comi-tragedy, that creek is called to this day, "Flour-bag Creek."

Now I take for granted that you are all more learned than these black fellows, and know quick-lime from flour. But still you are not bound to know what quick-lime is. Let me explain it to you.

Lime, properly speaking, is a metal, which goes among chemists by the name of calcium. But it is formed, as you all know, in the earth, not as a metal, but as a stone, as chalk or limestone, which is a carbonate of lime; that is, calcium combined with oxygen and carbonic-acid gases.

In that state it will make, if it is crystalline and hard, excellent building stone. The finest white marble, like that of Carrara in Italy, of which the most delicate statues are carved, is carbonate of lime altered and hardened by volcanic heat. But to make mortar of it, it must be softened and then brought into a state in which it can be hardened again; and ages since, some man or other, who deserves to rank as one of the great inventors, one of the great benefactors of his race, discovered the art of making lime soft and hard again; in fact of making mortar. The discovery was probably very ancient; and made, probably like most of the old discoveries, in the East, spreading Westward gradually. The earlier Greek buildings are cyclopean, that is, of stone fitted together without mortar. The earlier Egyptian buildings, though the stones are exquisitely squared and polished, are put together likewise without mortar. So, long ages after, were the earlier Roman buildings, and even some of the later. The famous aqueduct of the Pont du Gard, near Nismes, in the south of France, has, if I recollect right, no mortar whatever in it. The stones of its noble double tier of circular arches have been dropped into their places upon the wooden centres, and stand unmoved to this day, simply by the jamming of their own weight; a miracle of art. But the fact is puzzling; for these Romans were the best mortar makers of the world. We cannot, I believe, surpass them in the art even now; and in some of their old castles, the mortar is actually to this day harder and tougher than the stones which it holds together. And they had plenty of lime at hand if they had chosen to make mortar. The Pont du Gard crosses a limestone ravine, and is itself built of limestone. But I presume the cunning Romans would not trust mortar made from that coarse Nummulite limestone, filled with gritty sand, and preferred, with their usual carefulness, no mortar at all to bad.

But I must return, and tell my readers, in a few words, the chemical history of mortar. If limestone be burnt, or rather roasted, in a kiln, the carbonic acid is given off—as you may discover by your own nose; as many a poor tramp has discovered too late, when, on a cold winter night, he has lain down by the side of the burning kiln to keep himself warm, and woke in the other world, stifled to death by the poisonous fumes.

The lime then gives off its carbonic acid, and also its water of crystallisation, that is, water which it holds (as do many rocks) locked up in it unseen, and only to be discovered by chemical analysis. It is then anhydrous—that is, waterless—oxide of lime, what we call quick-lime; that which figured in the comitragedy of "Flour-bag Creek;" and then, as you may find if you get it under your nails or into your eyes, will burn and blister like an acid.

This has to be turned again into a hard and tough artificial limestone, in plain words, into mortar; and the first step is to slack it—that is, to give it back the water which it has lost, and for which it is as it were thirsting. So it is slacked with water, which it drinks in, heating itself and the water till it steams and swells in bulk, because it takes the substance of the water into its own substance. Slacked lime, as we all know, is not visibly wetter than quick-lime; it crumbles to a dry white powder in spite of all the water which it contains.

Then it must be made to set, that is, to return to limestone, to carbonate of lime, by drinking in the carbonic acid from water and air, which some sorts of lime will do instantly, setting at once, and being therefore used as cements. But the lime usually employed must be mixed with more or less sand to make it set hard: a mysterious process, of which it will be enough to tell the reader that the sand and lime are said to unite gradually, not only mechanically, that is, by sticking together; but also in part chemically—that is, by forming out of themselves a new substance, which is called silicate of lime.

Be that as it may, the mortar paste has now to do two things; first to dry, and next to take up carbonic acid from the air and water, enough to harden it again into limestone: and that it will take some time in doing. A thick wall, I am informed, requires several years before it is set throughout, and has acquired its full hardness, or rather toughness; and good mortar, as is well known, will acquire extreme hardness with age, probably from the very same cause that it did when it was limestone in the earth. For, as a general rule, the more ancient the strata is in which the limestone is found, the harder the limestone is; except in cases where volcanic action and earthquake pressure have hardened limestone in more recent strata, as in the case of the white marbles of Carrara in Italy, which are of the age of our Oolites, that is, of the freestone of Bath, etc., hardened by the heat of intruded volcanic rocks.

But now: what is the limestone? and how did it get where it is— not into the mortar, I mean, but into the limestone quarry? Let me tell you, or rather, help you to tell yourselves, by leading you, as before, from the known to the unknown. Let me lead you to places unknown indeed to most; but there may be sailors or soldiers among my readers who know them far better than I do. Let me lead you, in fancy, to some island in the Tropic seas. After all, I am not leading you as far away as you fancy by several thousand miles, as you will see, I trust, ere I have done.

Let me take you to some island: what shall it be like? Shall it be a high island, with cliff piled on cliff, and peak on peak, all rich with mighty forests, like a furred mantle of green velvet, mounting up and up till it is lost among white clouds above? Or shall it be a mere low reef, which you do not see till you are close upon it; on which nothing rises above the water, but here and there a knot of cocoa-nut palms or a block of stone, or a few bushes, swarming with innumerable sea-fowl and their eggs? Let it be which you will: both are strange enough; both beautiful; both will tell us a story.

The ship will have to lie-to, and anchor if she can; it may be a mile, it may be only a few yards, from the land. For between it and the land will be a line of breakers, raging in before the warm trade-wind. And this, you will be told, marks the edge of the coral reef.

You will have to go ashore in a boat, over a sea which looks unfathomable, and which may be a mile or more in depth, and search for an opening in the reef, through which the boat can pass without being knocked to pieces.

You find one: and in a moment, what a change! The deep has suddenly become shallow; the blue white, from the gleam of the white coral at the bottom. But the coral is not all white, only indeed a little of it; for as you look down through the clear water, you find that the coral is starred with innumerable live flowers, blue, crimson, grey, every conceivable hue; and that these are the coral polypes, each with its ring of arms thrust out of its cell, who are building up their common habitations of lime. If you want to understand, by a rough but correct description, what a coral polype is: all who have been to the sea-side know, or at least have heard of, sea-anemones. Now coral polypes are sea-anemones, which make each a shell of lime, growing with its growth. As for their shapes, the variety of them, the beauty of them, no tongue can describe them. If you want to see them, go to the Coral Rooms of the British or Liverpool Museums, and judge for yourselves. Only remember that you must re-clothe each of those exquisite forms with a coating of live jelly of some delicate hue, and put back into every one of the thousand cells its living flower; and into the beds, or rather banks, of the salt-water flower garden, the gaudiest of shell-less sea-anemones, such as we have on our coasts, rooted in the cracks, and live shells and sea-slugs, as gaudy as they, crawling about, with fifty other forms of fantastic and exuberant life. You must not overlook, too, the fish, especially the parrot-fish, some of them of the gaudiest colours, who spend their lives in browsing on the live coral, with strong clipping and grinding teeth, just as a cow browses the grass, keeping the animal matter, and throwing away the lime in the form of an impalpable white mud, which fills up the interstices in the coral beds.

The bottom, just outside the reef, is covered with that mud, mixed with more lime-mud, which the surge wears off the reef; and if you have, as you should have, a dredge on board, and try a haul of that mud as you row home, you may find, but not always, animal forms rooted in it, which will delight the soul of a scientific man. One, I hope, would be some sort of Terebratula, or shell akin to it. You would probably think it a cockle: but you would be wrong. The animal which dwells in it has about the same relationship to a cockle as a dog has to a bird. It is a Brachiopod; a family with which the ancient seas once swarmed, but which is rare now, all over the world, having been supplanted and driven out of the seas by newer and stronger forms of shelled animals. The nearest spot at which you are likely to dredge a live Brachiopod will be in the deep water of Loch Fyne, in Argyleshire, where two species still linger, fastened, strangely enough, to the smooth pebbles of a submerged glacier, formed in the open air during the age of ice, but sunk now to a depth of eighty fathoms. The first time I saw those shells come up in the dredge out of the dark and motionless abyss, I could sympathise with the feelings of mingled delight and awe which, so my companion told me, the great Professor Owen had in the same spot first beheld the same lingering remnants of a primæval world.

The other might be (but I cannot promise you even a chance of dredging that, unless you were off the coast of Portugal, or the windward side of some of the West India Islands) a live Crinoid; an exquisite starfish, with long and branching arms, but rooted in the mud by a long stalk, and that stalk throwing out barren side branches; the whole a living plant of stone. You may see in museums specimens of this family, now so rare, all but extinct. And yet fifty or a hundred different forms of the same type swarmed in the ancient seas: whole masses of limestone are made up of little else but the fragments of such animals.

But we have not landed yet on the dry part of the reef. Let us make for it, taking care meanwhile that we do not get our feet cut by the coral, or stung as by nettles by the coral insects. We shall see that the dry land is made up entirely of coral, ground and broken by the waves, and hurled inland by the storm, sometimes in huge boulders, mostly as fine mud;

and that, under the influence of the sun and of the rain, which filters through it, charged with lime from the rotting coral, the whole is setting, as cement sets, into rock. And what is this? A long bank of stone standing up as a low cliff, ten or twelve feet above high-water mark. It is full of fragments of shell, of fragments of coral, of all sorts of animal remains; and the lower part of it is quite hard rock. Moreover, it is bedded in regular layers, just such as you see in a quarry. But how did it get there? It must have been formed at the sea-level, some of it, indeed, under the sea; for here are great masses of madrepore and limestone corals imbedded just as they grew. What lifted it up? Your companions, if you have any who know the island, have no difficulty in telling you. It was hove up, they say, in the earthquake in such and such a year; and they will tell you, perhaps, that if you will go on shore to the main island which rises inside the reef, you may see dead coral beds just like these lying on the old rocks, and sloping up along the flanks of the mountains to several hundred feet above the sea. I have seen such many a time.

Thus you find the coral being converted gradually into a limestone rock, either fine and homogeneous, composed of coral grown into pulp, or filled with corals and shells, or with angular fragments of older coral rock. Did you never see that last? No? Yes, you have a hundred times. You have but to look at the marbles commonly used about these islands, with angular fragments imbedded in the mass, and here and there a shell, the whole cemented together by water holding in solution carbonate of lime, and there see the very same phenomenon perpetuated to this day.

Thus, I think, we have got first from the known to the unknown; from a tropic coral island back here to the limestone hills of Great Britain; and I did not speak at random when I said that I was not leading you away as far as you fancied by several thousand miles.

Examine any average limestone quarry from Bristol to Berwick, and you will see there all that I have been describing; that is, all of it which is not soft animal matter, certain to decay.

You will see the lime-mud hardened into rock beds; you will see the shells embedded in it; you will see the corals in every stage of destruction; you will see whole layers made up of innumerable fragments of Crinoids—no wonder they are innumerable, for, it has been calculated, there are in a single animal of some of the species 140,000 joints—140,000 bits of lime to fall apart when its soft parts decay. But is it not all there? And why should it not have got there by the same process by which similar old coral beds get up the mountain sides in the West Indies and elsewhere; namely, by the upheaving force of earthquakes? When you see similar effects, you have a right to presume similar causes. If you see a man fall off a house here, and break his neck; and some years after, in London or New York, or anywhere else, find another man lying at the foot of another house, with his neck broken in the same way, is it not a very fair presumption that he has fallen off a house likewise?

You may be wrong. He may have come to his end by a dozen other means: but you must have proof of that. You will have a full right, in science and in common sense, to say—That man fell off the house, till some one proves to you that he did not.

In fact, there is nothing which you see in the limestones of these isles—save and except the difference in every shell and coral—which you would not see in the coral-beds of the West Indies, if such earthquakes as that famous one at St. Thomas's, in 1866, became common and periodic, upheaving the land (they needs upheave it a very little, only two hundred and fifty feet), till St. Thomas's, and all the Virgin Isles, and the mighty mountain of Porto Rico, which looms up dim and purple to the west, were all joined into dry land once more, and the lonely coral-shoal of Anegada were raised, as it would be raised then, into a limestone table-land, like that of Central Ireland, of Galway, or of County Clare.

But you must clearly understand, that however much these coralline limestones have been upheaved since they were formed, yet the sea-bottom, while they were being formed, was sinking and not rising. This is a fact which was first pointed out by Mr. Darwin, from the observations which he made in the world-famous Voyage of the Beagle; and the observations of subsequent great naturalists have all gone to corroborate his theory.

It was supposed at first, you must understand, that when a coral island rose steeply to the surface of the sea out of blue water, perhaps a thousand fathoms or more, that fact was plain proof that the little coral polypes had begun at the bottom of the sea, and, in the course of ages, built up the whole island an enormous depth.

But it soon came out that that theory was not correct; for the coral polypes cannot live and build save in shallow water—say in thirty to forty fathoms. Indeed, some of the strongest and largest species work best at the very surface, and in the cut of the fiercest surf. And so arose a puzzle as to how coral rock is often found of vast thickness, which Mr. Darwin explained. His theory was, and there is no doubt now that it is correct, that in these cases the sea-bottom is sinking; that as it sinks, carrying the coral beds down with it, the coral dies, and a fresh live crop of polypes builds on the top of the houses of their dead ancestors: so that, as the depression goes on, generation after generation builds upwards, the living on the dead, keeping the upper surface of the reef at the same level, while its base is sinking downward into the abyss.

Applying this theory to the coral reef of the Pacific Ocean, the following interesting facts were made out:

That where you find an Island rising out of deep water, with a ring of coral round it, a little way from the shore—or, as in Eastern Australia, a coast with a fringing reef (the Flinders reef of Australia is eleven thousand miles long)—that is a pretty sure sign that that shore, or mountain, is sinking slowly beneath the sea. That where you find, as you often do in the Pacific, a mere atoll, or circular reef of coral, with a shallow pond of smooth water in the centre, and deep sea round, that is a pretty sure sign that the mountain-top has sunk completely into the sea, and that the corals are going on building where its peak once was.

And more. On working out the geography of the South Sea Islands by the light of this theory of Mr. Darwin's, the following extraordinary fact has been discovered:

That over a great part of the Pacific Ocean sinking is going on, and has been going on for ages; and that the greater number of the beautiful and precious South Sea Islands are only the remnants of a vast continent or archipelago, which once stretched for thousands of miles between Australia and South America.

Now, applying the same theory to limestone beds, which are, as you know, only fossil coral reefs, we have a right to say, when we see in England, Scotland, Ireland, limestones several thousand feet thick, that while they were being laid down as coral reef, the sea-bottom, and probably the neighbouring land, must have been sinking to the amount of their thickness—to several thousand feet—before that later sinking which enabled several hundred feet of millstone grit to be laid down on the top of the limestone.

This millstone grit is a new and a very remarkable element in our strange story. From Derby to Northumberland it forms vast and lofty moors, capping, as at Whernside and Penygent, the highest limestone hills with its hard, rough, barren, and unfossiliferous strata. Wherever it is found, it lies on the top of the "mountain," or carboniferous limestone. Almost everywhere, where coal is found in England, it lies on the millstone grit. I speak roughly, for fear of confusing my readers with details. The three deposits pass more or less, in many places, into each other: but always in the order of mountain limestone below, millstone grit on it, and coal on that again.

Now what does its presence prove? What but this? That after the great coral reefs which spread over Somersetshire and South Wales, around the present estuary of the Severn,—and those, once perhaps joined to them, which spread from Derby to Berwick, with a western branch through North-east Wales,—were laid down—after all this, I say, some change took place in the sea-bottom, and brought down on the reefs of coral sheets of sand, which killed the corals and buried them in grit. Does any reader wish for proof of this? Let him examine the "cherty," or flinty, beds which so often appear where the bottom of the millstone grit is passing into the top of the mountain limestone—the beds, to give an instance, which are now quarried on the top of the Halkin Mountain in Flintshire, for chert, which is sent to Staffordshire to be ground down for the manufacture of china. He will find layers in those beds, of several feet in thickness, as hard as flint, but as porous as sponge. On examining their cavities he will find them to be simply hollow casts of innumerable joints of Crinoids, so exquisitely preserved, even to their most delicate markings, that it is plain they were never washed about upon a beach, but have grown where, or nearly where, they lie. What then, has happened to them? They have been killed by the sand. The soft parts of the animals have decayed, letting the 140,000 joints (more or less) belonging to each animal fall into a heap, and be imbedded in the growing sand-rock; and then, it may be long years after, water filtering through the porous sand has removed the lime of which the joints were made, and left their perfect casts behind.

So much for the millstone grits. How long the deposition of sand went on, how long after it that second deposition of sands took place, which goes by the name of the "gannister," or lower coal-measures, we cannot tell. But it is clear, at least, that parts of that ancient sea were filling up and becoming dry land. For coal, or fossilised vegetable matter, becomes more and more common as we ascend in the series of beds; till at last, in the upper coal-measures, the enormous wealth of vegetation which grew, much of it, where it is now found, prove the existence of some such sheets of fertile and forest-clad lowland as I described in my last paper.

Thousands of feet of rich coral reef; thousands of feet of barren sands; then thousands of feet of rich alluvial forest—and all these sliding into each other, if not in one place, then in another, without violent break or change; this is the story which the lime in the mortar and the coal on the fire, between the two, reveal.

# VI. THE SLATES ON THE ROOF

The slates on the roof should be, when rightly understood, a pleasant subject for contemplation to the dweller in a town. I do not ask him to imitate the boy who, cliff-bred from his youth, used to spend stolen hours on the house-top, with his back against a chimney-stalk, transfiguring in his imagination the roof-slopes into mountain-sides, the slates into sheets of rock, the cats into lions, and the sparrows into eagles. I only wish that he should—at least after reading this paper—let the slates on the roof carry him back in fancy to the mountains whence they came; perhaps to pleasant trips to the lakes and hills of Cumberland, Westmoreland, and North Wales; and to recognise—as he will do if he have intellect as well as fancy—how beautiful and how curious an object is a common slate.

Beautiful, not only for the compactness and delicacy of its texture, and for the regularity and smoothness of its surface, but still more for its colour. Whether merely warm grey, as when dry, or bright purple, as when wet, the colour of the English slate well justifies Mr. Ruskin's saying, that wherever there is a brick wall and a slate roof there need be no want of rich colour in an English landscape. But most beautiful is the hue of slate, when, shining wet in the sunshine after a summer shower, its blue is brought out in rich contrast by golden spots of circular lichen, whose spores, I presume, have travelled with it off its native mountains. Then, indeed, it reminds the voyager of a sight which it almost rivals in brilliancy—of the sapphire of the deep ocean, brought out into blazing intensity by the contrast of the golden patches of floating gulf-weed beneath the tropic sun.

Beautiful, I say, is the slate; and curious likewise, nay, venerable; a most ancient and elaborate work of God, which has lasted long enough, and endured enough likewise, to bring out in it whatsoever latent capabilities of strength and usefulness might lie hid in it; which has literally been—as far as such words can apply to a thing inanimate—

> *Heated hot with burning fears,*
> *And bathed in baths of hissing tears,*
> *And battered by the strokes of doom*
> *To shape and use.*

And yet it was at first naught but an ugly lump of soft and shapeless ooze.

Therefore, the slates to me are as a parable, on which I will not enlarge, but will leave each reader to interpret it for himself. I shall confine myself now to proofs that slate is hardened mud, and to hints as to how it assumed its present form.

That slate may have been once mud, is made probable by the simple fact that it can be turned into mud again. If you grind tip slate, and then analyse it, you will find its mineral constituents to be exactly those of a fine, rich, and tenacious clay. The slate districts (at least in Snowdon) carry such a rich clay on them, wherever it is not masked by the ruins of other rocks. At Ilfracombe, in North Devon, the passage from slate below to clay above, may be clearly seen. Wherever the top of the slate beds, and the soil upon it, is laid bare, the black layers of slate may be seen gradually melting—if I may use the word—under the influence of rain and frost, into a rich tenacious clay, which is now not black, like its parent slate, but red, from the oxidation of the iron which it contains.

But, granting this, how did the first change take place?

It must be allowed, at starting, that time enough has elapsed, and events enough have happened, since our supposed mud began first to become slate, to allow of many and strange transformations. For these slates are found in the oldest beds of rocks, save one series, in the known world; and it is notorious that the older and lower the beds in which the slates are found, the better, that is, the more perfectly elaborate, is the slate. The best slates of Snowdon—I must confine myself to the district which I know personally—are found in the so-called "Cambrian" beds. Below these beds but one series of beds is as yet known in the world, called the "Laurentian." They occur, to a thickness of some eighty thousand feet, in Labrador, Canada, and the Adirondack mountains of New York: but their representatives in Europe are, as far as is known only to be found in the north-west highlands of Scotland, and in the island of Lewis, which consists entirely of them. And it is to be remembered, as a proof of their inconceivable antiquity, that they have been upheaved and shifted long before the Cambrian rocks were laid down "unconformably" on their worn and broken edges.

Above the "Cambrian" slates—whether the lower and older ones of Penrhyn and Llanberris, which are the same—one slate mountain being worked at both sides in two opposite valleys—or the upper and newer slates of Tremadoc, lie other and newer slate-bearing beds of inferior quality, and belonging to a yet newer world, the "Silurian." To them belong the Llandeilo flags and slates of Wales, and the Skiddaw slates of Cumberland, amid beds abounding in extinct fossil forms. Fossil shells are found, it is true, in the upper Cambrian beds. In the lower they have all but disappeared. Whether their traces have been obliterated by heat and pressure, and chemical action, during long ages; or whether, in these lower beds, we are actually reaching that "Primordial Zone" conceived of by M. Barrande, namely, rocks which existed before living things had begun to people this planet, is a question not yet answered. I believe the former theory to be the true one. That there was life, in the sea at least, even before the oldest Cambrian rocks were laid down, is proved by the discovery of the now famous fossil, the

Eozoon, in the Laurentian limestones, which seems to have grown layer after layer, and to have formed reefs of limestone as do the living coral-building polypes. We know no more as yet. But all that we do know points downwards, downwards still, warning us that we must dig deeper than we have dug as yet, before we reach the graves of the first living things.

Let this suffice at present for the Cambrian and Laurentian rocks.

The Silurian rocks, lower and upper, which in these islands have their chief development in Wales, and which are nearly thirty-eight thousand feet thick; and the Devonian or Old Red sandstone beds, which in the Fans of Brecon and Carmarthen-shire attain a thickness of ten thousand feet, must be passed through in an upward direction before we reach the bottom of that Carboniferous Limestone of which I spoke in my last paper. We thus find on the Cambrian rocks forty-five thousand feet at least of newer rocks, in several cases lying unconformably on each other, showing thereby that the lower beds had been up-heaved, and their edges worn off on a sea-shore, ere the upper were laid down on them; and throughout this vast thickness of rocks, the remains of hundreds of forms of animals, corals, shells, fish, older forms dying out in the newer rocks, and new ones taking their places in a steady succession of ever-varying forms, till those in the upper beds have become unlike those in the lower, and all are from the beginning more or less unlike any existing now on earth. Whole families, indeed, disappear entirely, like the Trilobites, which seem to have swarmed in the Silurian seas, holding the same place there as crabs and shrimps do in our modern seas. They vanish after the period of the coal, and their place is taken by an allied family of Crus-taceans, of which only one form (as far as I am aware) lingers now on earth, namely, the "King Crab," or Limulus, of the Indian Seas, a well-known animal, of which specimens may sometimes be seen alive in English aquaria. So perished in the lapse of those same ages, the armour-plated or "Ganoid" fish which Hugh Miller made so justly famous—and which made him so justly famous in return—appearing first in the upper Silurian beds, and abounding in vast variety of strange forms in the old Red Sandstone, but gradually disappearing from the waters of the world, till their only representatives, as far as known, are the Lepidostei, or "Bony Pikes," of North America; the Polypteri of the Nile and Senegal; the Lepidosirens of the African lakes and Western rivers; the Ceratodus or Barramundi of Queensland (the two latter of which approach Amphibians), and one or two more fantastic forms, either rudimentary or degraded, which have lasted on here and there in isolated stations through long ages, comparatively unchanged while all the world is changed around them, and their own kindred, buried like the fossil Ceratodus of the Trias beneath thousands of feet of ancient rock, among creatures the likes whereof are not to be found now on earth. And these are but two examples out of hundreds of the vast changes which have taken place in the animal life of the globe, between the laying down of the Cambrian slates and the present time.

Surely—and it is to this conclusion I have been tending throughout a seemingly wandering paragraph—surely there has been time enough during all those ages for clay to change into slate.

And how were they changed?

I think I cannot teach my readers this more simply than by asking them first to buy Sheet No. LXXVIII. S.E. (Bangor) of the Snowdon district of the Government Geological Survey, which may be ordered at any good stationer's, price 3s.; and study it with me. He will see down the right-hand margin interpretations of the different colours which mark the different beds, beginning with the youngest (alluvium) atop, and going down through Carboniferous Limestone and Sandstone, Upper Silurian, Lower Silurian, Cambrian, and below them certain rocks marked of different shades of red, which signify rocks either altered by heat, or poured out of old volcanic vents. He will next see that the map is covered with a labyrinth of red patches and curved lines, signifying the outcrop or appearance at the surface of these volcanic beds. They lie at every conceivable slope; and the hills and valleys have been scooped out by rain and ice into every conceivable slope likewise. Wherefore we see, here a broad patch of red, where the back of a sheet of Lava, Porphyry, Greenstone, or what not is exposed; there a narrow line curving often with the curve of the hill-side, where only the edge of a similar sheet is exposed; and every possible variety of shape and attitude between these two. He will see also large spaces covered with little coloured dots, which signify (as he will find at the margin) beds of volcanic ash. If he look below the little coloured squares on the margin, he will see figures marking the strike, or direction of the inclination of the beds— inclined, vertical, horizontal, contorted; that the white lines in the map signify faults, i.e. shifts in the strata; the gold lines, lodes of metal—the latter of which I should advise him strong- ly, in this district at least, not to meddle with: but to button up his pockets, and to put into the fire, in wholesome fear of his own weakness and ignorance, any puffs of mining companies which may be sent him—as one or two have probably been sent him already.

Furnished with which keys to the map, let him begin to con it over, sure that there is if not an order, still a grand meaning in all its seeming confusion; and let him, if he be a courteous and grateful person, return due thanks to Professor Ramsay for having found it all out; not without wondering, as I have often wondered, how even Professor Ramsay's acuteness and industry could find it all out.

When my reader has studied awhile the confusion—for it is a true confusion—of the different beds, he will ask, or at least have a right to ask, what known process of nature can have produced it? How have these various volcanic rocks, which he sees marked as Felspathic Traps, Quartz Porphyries, Green- stones, and so forth, got intermingled with beds which he is told to believe are volcanic ashes, and those again with fossil- bearing Silurian beds and Cambrian slates, which he is told to believe were deposited under water? And his puzzle will not be lessened when he is told that, in some cases, as in that of the summit of Snowdon, these very volcanic ashes contain fossil shells.

The best answer I can give is to ask him to use his imagination, or his common sense; and to picture to himself what must go on in the case of a submarine eruption, such as broke out off the coast of Iceland in 1783 and 1830, off the Azores in 1811, and in our day in more than one spot in the Pacific Ocean.

A main bore or vent—or more than one—opens itself between the bottom of the sea and the nether fires. From each rushes an enormous jet of high-pressure steam and other gases, which boils up through the sea, and forms a cloud above; that cloud descends again in heavy rain, and gives out often true lightning from its under side.

But it does more. It acts as a true steam-gun, hurling into the air fragments of cold rock rasped off from the sides of the bore, and fragments also of melted lava, and clouds of dust, which fall again into the sea, and form there beds either of fine mud or of breccia—that is, fragments of stone embedded in paste. This, the reader will understand, is no fancy sketch, as far as I am concerned. I have steamed into craters sawn through by the sea, and showing sections of beds of ash dipping outwards and under the sea, and in them boulders and pebbles of every size, which had been hurled out of the crater; and in them also veins of hardened lava, which had burrowed out through the soft ashes of the cone. Of those lava veins I will speak presently. What I want the reader to think of now is the immense quantity of ash which the steam-mitrailleuse hurls to so vast a height into the air, that it is often drifted many miles down to leeward. To give two instances: The jet of steam from Vesuvius, in the eruption of 1822, rose more than four miles into the air; the jet from the Souffrière of St. Vincent in the West Indies, in 1812, probably rose higher; certainly it met the N.E. trade-wind, for it poured down a layer of ashes, several inches thick, not only on St. Vincent itself, but on Barbadoes, eighty miles to windward, and therefore on all the sea between. Now let us consider what that represents—a layer of fine mud, laid down at the bottom of the ocean, several inches thick, eighty miles at least long, and twenty miles perhaps broad, by a single eruption. Suppose that hardened in long ages (as it would be under pressure) into a bed of fine grained Felstone, or volcanic ash; and we can understand how the ash-beds of Snowdonia—

which may be traced some of them for many square miles—
were laid down at the bottom of an ancient sea.

But now about the lavas or true volcanic rocks, which are
painted (as is usual in geological maps) red. Let us go down to
the bottom of the sea, and build up our volcano towards the
surface.

First, as I said, the subterranean steam would blast a bore.
The dust and stones, rasped and blasted out of that hole would
be spread about the sea-bottom as an ash-bed sloping away
round the hole; then the molten lava would rise in the bore,
and flow out over the ashes and the sea-bottom—perhaps in
one direction, perhaps all round. Then, usually, the volcano,
having vented itself, would be quieter for a time, till the heat
accumulated below, and more ash was blasted out, making
a second ash-bed; and then would follow a second lava flow.
Thus are produced the alternate beds of lava and ash which
are so common.

Now suppose that at this point the volcano was exhausted, and
lay quiet for a few hundred years, or more. If there was any
land near, from which mud and sand were washed down, we
might have layers on layers of sediment deposited, with live
shells, etc., living in them, which would be converted into fossils when they died; and so we should have fossiliferous beds
over the ashes and lavas. Indeed, shells might live and thrive
in the ash-mud itself, when it cooled, and the sea grew quiet,
as they have lived and thriven in Snowdonia.

Now suppose that after these sedimentary beds are laid down
by water, the volcano breaks out again—what would happen?

Many things: specially this, which has often happened already.

The lava, kept down by the weight of these new rocks, searches
for the point of least resistance, and finds it in a more horizontal direction. It burrows out through the softer ash-beds, and
between the sedimentary beds, spreading itself along horizontally. This process accounts for the very puzzling, though
very common case in Snowdon and elsewhere, in which we

find lavas interstratified with rocks which are plainly older than those lavas. Perhaps when that is done the volcano has got rid of all its lava, and is quiet. But if not, sooner or later, it bores up through the new sedimentary rocks, faulting them by earthquake shocks till it gets free vent, and begins its layers of alternate ash and lava once more.

And consider this fact also: If near the first (as often happens) there is another volcano, the lava from one may run over the lava from the other, and we may have two lavas of different materials overlying each other, which have come from different directions. The ashes blown out of the two craters may mingle also, and so, in the course of ages, the result may be such a confusion of ashes, lavas, and sedimentary rocks as we find throughout most mountain ranges in Snowdon, in the Lake mountains, in the Auvergne in France, in Sicily round Etna, in Italy round Vesuvius, and in so many West Indian Islands; the last confusion of which is very likely to be this:

That when the volcano has succeeded—as it did in the case of Sabrina Island off the Azores in 1811, and as it did, perhaps often, in Snowdonia—in piling up an ash cone some hundred feet out of the sea; that—as has happened to Sabrina Island—the cone is sunk again by earthquakes, and gnawn down at the same time by the sea-waves, till nothing is left but a shoal under water. But where have all its vast heaps of ashes gone? To be spread about over the bottom of the sea, to mingle with the mud already there, and so make beds of which, like many in Snowdon, we cannot say whether they are of volcanic or of marine origin, because they are of both.

But what has all this to do with the slates?

I shall not be surprised if my readers ask that question two or three times during this paper. But they must be kind enough to let me tell my story my own way. The slates were not made in a day, and I fear they cannot be explained in an hour: unless we begin carefully at the beginning in order to end at the end. Let me first make my readers clearly understand that all our slate-bearing mountains, and most also of the non-slate-bearing ones likewise, are formed after the fashion which I have described, namely, beneath the sea. I do not say that there may not have been, again, and again, ash-cones rising above the surface of the waves. But if so, they were washed away, again and again, ages before the land assumed anything of its present shape; ages before the beds were twisted and upheaved as they are now.

And therefore I beg my readers to put out of their minds once and for all the fancy that in any known part of these islands craters are to be still seen, such as exist in Etna, or Vesuvius, or other volcanoes now at work in the open air.

It is necessary to insist on this, because many people hearing that certain mountains are volcanic, conclude—and very naturally and harmlessly—that the circular lakes about their tops are true craters. I have been told, for instance, that that wonderful little blue Glas Llyn, under the highest cliff of Snowdon, is the old crater of the mountain; and I have heard people insist that a similar lake, of almost equal grandeur, in the south side of Cader Idris, is a crater likewise.

But the fact is not so. Any one acquainted with recent craters would see at once that Glas Llyn is not an ancient one; and I am not surprised to find the Government geologists declaring that the Llyn on Cader Idris is not one either. The fact is, that the crater, or rather the place where the crater has been, in ancient volcanoes of this kind, is probably now covered by one of the innumerable bosses of lava.

For, as an eruption ceases, the melted lava cools in the vents, and hardens; usually into lava infinitely harder than the ash-cone round it; and this, when the ash-cone is washed off, remains as the highest part of the hill, as in the Mont Dore and the Cantal in France, and in several extinct volcanoes in the Antilles. Of course the lava must have been poured out, and the ashes blown out from some vents or other, connected with the nether world of fire; probably from many successive vents. For in volcanoes, when one vent is choked, another is wont to open at some fresh point of least resistance among the overlying rocks. But where are these vents? Buried deep under successive eruptions, shifted probably from their places by successive upheavings and dislocations; and if we wanted to find them we should have to quarry the mountain range all over, a mile deep, before we hit upon here and there a tap-root of ancient lava, connecting the upper and the nether worlds. There are such tap-roots, probably, under each of our British mountain ranges. But Snowdon, certainly, does not owe its shape to the fact of one of these old fire vents being under it. It owes its shape simply to the accident of some of the beds toward the summit being especially hard, and thus able to stand the wear and tear of sea-wave, ice, and rain. Its lakes have been formed quite regardless of the lie of the rocks, though not regardless of their relative hardness. But what forces scooped them out—whether they were originally holes left in the ground by earthquakes, and deepened since by rain and rivers, or whether they were scooped out by ice, or by any other means, is a question on which the best geologists are yet undecided—decided only on this—that craters they are not.

As for the enormous changes which have taken place in the outline of the whole of the mountains, since first their strata were laid down at the bottom of the sea: I shall give facts enough, before this paper is done, to enable readers to judge of them for themselves.

The reader will now ask, naturally enough, how such a heap of beds as I have described can take the shape of mountains like Snowdon.

Look at any sea cliff in which the strata are twisted and set on slope. There are hundreds of such in these isles. The beds must have been at one time straight and horizontal. But it is equally clear that they have been folded by being squeezed laterally. At least, that is the simplest explanation, as may be proved by experiment. Take a number of pieces of cloth, or any such stuff; lay them on each other and then squeeze them together at each end. They will arrange themselves in folds, just as the beds of the cliff have done. And if, instead of cloth, you take some more brittle matter, you will find that, as you squeeze on, these folds will tend to snap at the points of greatest tension or stretching, which will be of course at the anticlinal and synclinal lines—in plain English, the tops and bottoms of the folds. Thus cracks will be formed; and if the pressure goes on, the ends of the layers will shift against each other in the line of those cracks, forming faults like those so common in rocks.

But again, suppose that instead of squeezing these broken and folded lines together any more, you took off the pressure right and left, and pressed them upwards from below, by a mimic earthquake. They would rise; and as they rose leave open space between them. Now if you could contrive to squeeze into them from below a paste, which would harden in the cracks and between the layers, and so keep them permanently apart, you would make them into a fair likeness of an average mountain range—a mess—if I may make use of a plain old word—of rocks which have, by alternate contraction and expansion, helped in the latter case by the injection of molten lava, been thrust about as they are in most mountain ranges.

That such a contraction and expansion goes on in the crust of the earth is evident; for here are the palpable effects of it. And the simplest general cause which I can give for it is this: That things expand as they are heated, and contract as they are cooled.

Now I am not learned enough—and were I, I have not time—to enter into the various theories which philosophers have put forward, to account for these grand phenomena.

The most remarkable, perhaps, and the most probable, is the theory of M. Elie de Beaumont, which is, in a few words, this:

That this earth, like all the planets, must have been once in a state of intense heat throughout, as its mass inside is probably now.

That it must be cooling, and giving off its heat into space.

That, therefore, as it cools, its crust must contract.

That, therefore, in contracting, wrinkles (for the loftiest mountain chains are nothing but tiny wrinkles, compared with the whole mass of the earth), wrinkles, I say, must form on its surface from time to time. And that the mountain chains are these wrinkles.

Be that as it may, we may safely say this. That wherever the internal heat of the earth tends (as in the case of volcanoes) towards a particular spot, that spot must expand, and swell up, bulging the rocks out, and probably cracking them, and inserting melting lava into those cracks from below. On the other hand, if the internal heat leaves that spot again, and it cools, then it must contract more or less, in falling inward toward the centre of the earth; and so the beds must be crumpled, and crushed, and shifted against each other still more, as those of our mountains have been.

But here may arise, in some of my readers' minds, a reasonable question—If these upheaved beds were once horizontal, should we not be likely to find them, in some places, horizontal still?

A reasonable question, and one which admits of a full answer.

They know, of course, that there has been a gradual, but steady, change in the animals of this planet; and that the relative age of beds can, on the strength of that known change, be determined generally by the fossils, usually shells, peculiar to them: so that if we find the same fashion of shells, and still more the same species of shells, in two beds in different quarters of the world, then we have a right to say—These beds were laid down at least about the same time. That is a general rule among all geologists, and not to be gainsaid.

Now I think I may say, that, granting that we can recognise a bed by its fossils, there are few or no beds which are found in one place upheaved, broken, and altered by heat, which are not found in some other place still horizontal, unbroken, unaltered, and more or less as they were at first.

From the most recent beds; from the upheaved coral-rocks of the West Indies, and the upheaved and faulted boulder clay and chalk of the Isle of Moen in Denmark—downwards through all the strata, down to that very ancient one in which the best slates are found, this rule, I believe, stands true.

It stands true, certainly, of the ancient Silurian rocks of Wales, Cumberland, Ireland, and Scotland.

For, throughout great tracts of Russia, and in parts of Norway and Sweden, Sir Roderick Murchison discovered our own Silurian beds, recognisable from their peculiar fossils. But in what state? Not contracted, upheaved, and hardened to slates and grits, as they are in Wales and elsewhere: but horizontal, unbroken, and still soft, because undisturbed by volcanic rooks and earthquakes. At the bottom of them all, near Petersburg, Sir Roderick found a shale of dried mud (to quote his own words), "so soft and incoherent that it is even used by sculptors for modelling, although it underlies the great mass of fossil-bearing Silurian rocks, and is, therefore, of the same age as the lower crystalline hard slates of North Wales. So entirely have most of these eldest rocks in Russia been exempted from the influence of change, throughout those enormous periods which have passed away since their accumulation."

Among the many discoveries which science owes to that illustrious veteran, I know none more valuable for its bearing on the whole question of the making of the earth-crust, than this one magnificent fact.

But what a contrast between these Scandinavian and Russian rocks and those of Britain! Never exceeding, in Scandinavia, a thousand feet in thickness, and lying usually horizontal, as they were first laid down, they are swelled in Britain to a thickness of thirty thousand feet, by intruded lavas and ashes; snapt, turned, set on end at every conceivable angle; shifted against each other to such an extent, that, to give a single instance, in the Vale of Gwynnant, under Snowdon, an immense wedge of porphyry has been thrust up, in what is now the bottom of the valley, between rocks far newer than it, on one side to a height of eight hundred, on the other to a height of eighteen hundred feet—half the present height of Snowdon. Nay, the very slate beds of Snowdonia have not forced their way up from under the mountain—without long and fearful struggles. They are set in places upright on end, then horizontal again, then sunk in an opposite direction, then curled like sea-waves, then set nearly upright once more, and faulted through and through, six times, I believe, in the distance of a mile or two; they carry here and there on their backs patches of newer beds, the rest of which has long vanished; and in their rise they have hurled back to the eastward, and set upright, what is now the whole western flank of Snowdon, a mass of rock which was then several times as thick as it is now.

The force which thus tortured them was probably exerted by the great mass of volcanic Quartz-porphyry, which rises from under them to the north-west, crossing the end of the lower lake of the Llanberris; and indeed the shifts and convulsions which have taken place between them and the Menai Straits are so vast that they can only be estimated by looking at them on the section which may be found at the end of Professor Ramsay's "Geological Survey of North Wales." But anyone who will study that section, and use (as with the map) a little imagination and common sense, will see that between the heat of that Porphyry, which must have been poured out as a fluid mass as hot, probably, as melted iron, and the pressure of it below, and of the Silurian beds above, the Cambrian mud-strata of Llanberris and Penrhyn quarries must have suffered enough to change them into something very different from mud, and, therefore, probably, into what they are now—namely, slate.

And now, at last, we have got to the slates on the roof, and may disport ourselves over them—like the cats.

Look at any piece of slate. All know that slate splits or cleaves freely, in one direction only, into flat layers. Now any one would suppose at first sight, and fairly enough, that the flat surface—the "plane of cleavage"—was also the plane of bedding. In simpler English we should say—The mud which has hardened into the slate was laid down horizontally; and therefore each slate is one of the little horizontal beds of it, perhaps just what was laid down in a single tide. We should have a right to do so, because that would be true of most sedimentary rocks. But it would not be true of slate. The plane of bedding in slate has nothing to do with the plane of cleavage. Or, more plainly, the mud of which the slate is made may have been deposited at the sea-bottom at any angle to the plane of cleavage. We may sometimes see the lines of the true bedding—the lines which were actually horizontal when the mud was laid down—in bits of slate, and find them sometimes perpendicular to, sometimes inclined to, and sometimes again coinciding with the plane of cleavage, which they have evidently acquired long after.

Nay, more. These parallel planes of cleavage, at each of which the slate splits freely, will run through a whole mountain at the same angle, though the beds through which they run may be tilted at different angles, and twisted into curves.

Now what has made this change in the rook? We do not exactly know. One thing is clear, that the particles of the now solid rock have actually moved on themselves. And this is proved by a very curious fact—which the reader, if he geologises about slate quarries much, may see with his own eyes. The fossils in the slate are often distorted into quaint shapes, pulled out long if they lie along the plane of cleavage, or squeezed together, or doubled down on both sides, if they lie across the plane. So that some force has been at work which could actually change the shape of hard shells, very slowly, no doubt, else it would have snapped and crumbled them.

If I am asked what that force was, I do not know. I should advise young geologists to read what Sir Henry de la Beche has said on it in his admirable "Geological Observer," pp. 706-725. He will find there, too, some remarks on that equally mysterious phenomena of jointing, which you may see in almost all the older rocks; it is common in limestones. All we can say is, that some force has gone on, or may be even now going on, in the more ancient rocks, which is similar to that which produces single crystals; and similar, too, to that which produced the jointed crystals of basalt, *i.e.* lava, at the Giant's Causeway, in Ireland, and Staffa, in the Hebrides. Two philosophers—Mr. Robert Were Fox and Mr. Robert Hunt—are of opinion that the force which has determined the cleavage of slates may be that of the electric currents, which (as is well known) run through the crust of the earth. Mr. Sharpe, I believe, attributes the cleavage to the mere mechanical pressure of enormous weights of rock, especially where crushed by earthquakes. Professor Rogers, again, points out that as these slates may have been highly heated, thermal electricity (*i.e.* electricity brought out by heat) may have acted on them.

One thing at least is clear. That the best slates are found among ancient lavas, and also in rocks which are faulted and tilted enormously, all which could not have happened without a proportionately enormous pressure, and therefore heat; and next, that the best slates are invariably found in the oldest beds— that is, in the beds which have had most time to endure the changes, whether mechanical or chemical, which have made the earth's surface what we see it now.

Another startling fact the section of Snowdonia, and I believe of most mountain chains in these islands, would prove— namely, that the contour of the earth's surface, as we see it now, depends very little, certainly in mountains composed of these elder rocks upon the lie of the strata, or beds, but has been carved out by great forces, long after those beds were not only laid down and hardened, but faulted and tilted on end. Snowdon itself is so remarkable an instance of this fact that, as it is a mountain which every one in these happy days of excursion-trains and steamers either has seen or can see, I must say a few more words about it.

Any one who saw that noble peak leaping high into the air, dominating all the country round, at least upon three sides, and was told that its summit consisted of beds much newer, not much older, than the slate-beds fifteen hundred feet down on its north-western flank—any one, I say, would have the right at first sight, on hearing of earthquake faults and upheavals, to say—The peak of Snowdon has been upheaved to its present height above and out of the lower lands around. But when he came to examine sections, he would find his reasonable guess utterly wrong. Snowdon is no swelling up of the earth's crust.

The beds do not, as they would in that case, slope up to it. They slope up from it, to the north-west in one direction, and the south-south-west in the other; and Snowdon is a mere insignificant boss, left hanging on one slope of what was once an enormous trough, or valley, of strata far older than itself. By restoring these strata, in the direction of the angles, in which they crop out, and vanish at the surface, it is found that to the north-west—the direction of the Menai Straits—they must once have risen to a height of at least six or seven thousand feet; and more, by restoring them, specially the ash-bed of Snowdon, towards the south-east—which can be done by the guidance of certain patches of it left on other hills—it is found that south of Ffestiniog, where the Cambrian rocks rise again to the surface, the south side of the trough must have sloped upwards to a height of from fifteen to twenty thousand feet, whether at the bottom of the sea, or in the upper air, we cannot tell. But the fact is certain, that off the surface of Wales, south of Ffestiniog a mass of solid rock as high as the Andes has been worn down and carried bodily away; and that a few miles south again, the peak of Arran Mowddy, which is now not two thousand feet high, was once—either under the sea or above it—nearer ten thousand feet.

If I am asked whither is all that enormous mass of rock—millions of tons—gone? Where is it now? I know not. But if I dared to hazard a guess, I should say it went to make the New Red sandstones of England.

The New Red sandstones must have come from somewhere. The most likely region for them to have come from is from North Wales, where, as we know, vast masses of gritty rock have been ground off, such as would make fine sandstones if they had the chance. So that many a grain of sand in Chester walls was probably once blasted out of the bowels of the earth into the old Silurian sea, and after a few hundreds of thousands of years' repose in a Snowdonian ash-bed, was sent eastward to build the good old city and many a good town more.

And the red marl—the great deposit of red marl which covers a wide region of England—why should not it have come from the same quarter? Why should it not be simply the remains of the Snowdon Slate? Mud the slate was, and into mud it has returned. Why not? Some of the richest red marl land I know, is, as I have said, actually being made now, out of the black slates of Ilfracombe, wherever they are weathered by rain and air. The chemical composition is the same. The difference in colour between black slate and red marl is caused simply by the oxidation of the iron in the slate.

And if my readers want a probable cause why the sandstones lie undermost, and the red marl uppermost—can they not find one for themselves? I do not say that it is the cause, but it is at least a causa vera, one which would fully explain the fact, though it may be explicable in other ways. Think, then, or shall I think for my readers?

Then do they not see that when the Welsh mountains were ground down, the Silurian strata, being uppermost, would be ground down first, and would go to make the lower strata of the great New Red Sandstone Lowland; and that being sandy, they would make the sandstones? But wherever they were ground through, the Lower Cambrian slates would be laid bare; and their remains, being washed away by the sea the last, would be washed on to the top of the remains of the Silurians; and so (as in most cases) the remains of the older rock, when redeposited by water, would lie on the remains of the younger rock. And do they not see that (if what I just said is true) these slates would grind up into red marl, such as is seen over the west and south of Cheshire and Staffordshire and far away into Nottinghamshire? The red marl must almost certainly have been black slate somewhere, somewhen. Why should it not have been such in Snowdon? And why should not the slates in the roof be the remnants of the very beds which are now the marl in the fields?

And thus I end my story of the slates in the roof, and these papers on Town Geology. I do so, well knowing how imperfect they are: though not, I believe, inaccurate. They are, after all, merely suggestive of the great amount that there is to be learnt about the face of the earth and how it got made, even by the townsman, who can escape into the country and exchange the world of man for the world of God, only, perhaps, on Sundays—if, alas! even then—or only once a year by a trip in a steamer or an excursion train. Little, indeed, can he learn of the planet on which he lives. Little in that direction is given to him, and of him little shall be required. But to him, for that very reason, all that can be given should be given; he should have every facility for learning what he can about this earth, its composition, its capabilities; lest his intellect, crushed and fettered by that artificial drudgery which we for a time miscall civilisation, should begin to fancy, as too many do already, that the world is composed mainly of bricks and deal, and governed by acts of parliament. If I shall have awakened any townsmen here and there to think seriously of the complexity, the antiquity, the grandeur, the true poetry, of the commonest objects around them, even the stones beneath their feet; if I shall have suggested to them the solemn thought that all these things, and they themselves still more, are ordered by laws, utterly independent of man's will about them, man's belief in them; if I shall at all have helped to open their eyes that they may see, and their ears that they may hear, the great book which is free to all alike, to peasant as to peer, to men of business as to men of science, even that great book of nature, which is, as Lord Bacon said of old, the Word of God revealed in facts—then I shall have a fresh reason for loving that science of geology, which has been my favourite study since I was a boy.

LaVergne, TN USA
30 November 2009

165579LV00010B/301/P